BOOK 1

NAME:

CLASS:

NELSON MATHS

WORKBOOK

YEARS 7–10

Kuldip Khehra
Judy Binns
Gaspare Carrozza
Robert Yen
Sandra Tisdell-Clifford

HOMEWORK ASSIGNMENTS

Nelson Maths Workbook 1
1st Edition
Kuldip Khehra
Judy Binns
Gaspare Carrozza
Robert Yen
Sandra Tisdell-Clifford
ISBN 9780170454452

Publisher: Robert Yen
Project editor: Alan Stewart
Editor: Anna Pang
Cover design: James Steer
Original text design by Alba Design, Adapted by: James Steer
Cover image: iStockphoto/iam-Citrus
Project design: James Steer
Illustrator: Cat MacInnes
Typeset by: MPS Limited
Production controller: Karen Young

For product information and technology assistance,
in Australia call **1300 790 853**;
in New Zealand call **0800 449 725**

For permission to use material from this text or product, please email **aust.permissions@cengage.com**

ISBN 978 0 17 045445 2

Cengage Learning Australia
Level 7, 80 Dorcas Street
South Melbourne, Victoria Australia 3205

Cengage Learning New Zealand
Unit 4B Rosedale Office Park
331 Rosedale Road, Albany, North Shore 0632, NZ

For learning solutions, visit **cengage.com.au**

Printed in China by 1010 Printing International Limited.
2 3 4 5 6 7 26 25 24 23 22

This 200-page workbook contains worksheets, puzzles, StartUp assignments and homework assignments written for the Australian Curriculum in Mathematics. It can be used as a valuable resource for teaching Year 7 mathematics, regardless of the textbook used in the classroom, and takes a wholistic approach to the curriculum, including some Year 6 and Year 8 work as well. This workbook is designed to be handy for homework, assessment, practice, revision, relief classes or 'catch-up' lessons.

Inside:

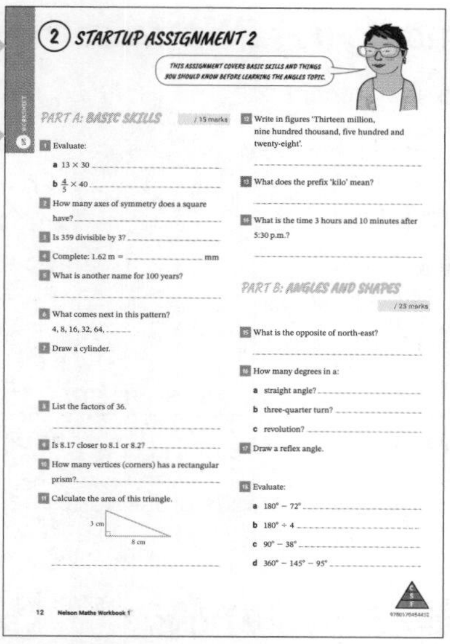

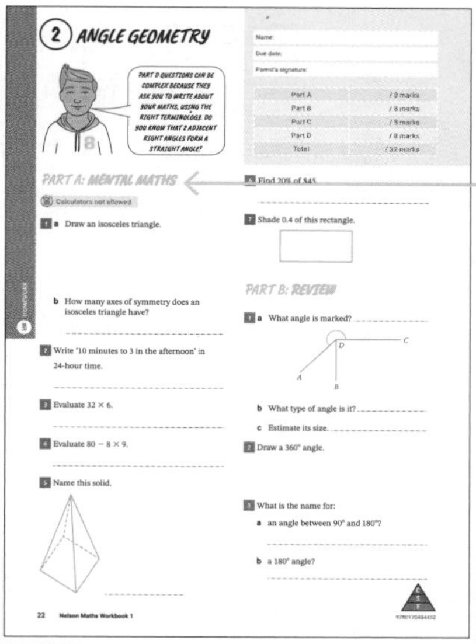

WS WORKSHEETS ordered by topic providing further practice, application, remedial and revision work

PS PUZZLE SHEETS containing matching games, codes and puzzles

Weekly **HW HOMEWORK** assignments that include **Mental maths** (calculator-free), Topic **Review** and **Practice**, including **Numeracy and literacy** skills

StartUp assignments beginning each topic, revising skills from previous topics and prerequisite knowledge for the topic, including basic skills, review of a specific topic and a challenge problem

Word puzzles, such as a crossword or find-a-word, that reinforce the language of mathematics learned in the topic

The ideas and activities presented in this book were written by practising teachers and used successfully in the classroom.

Colour-coding of selected questions

Questions on most worksheets are graded by level of difficulty:

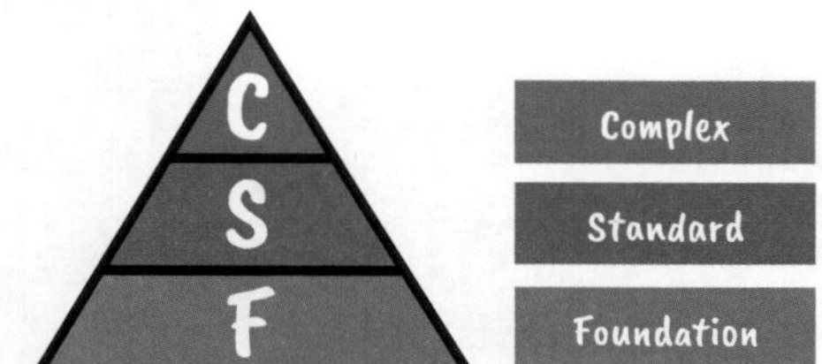

CONTENTS

 WORKSHEET  PUZZLE SHEET  HOMEWORK

MEET YOUR MATHS GUIDES ...

INTRODUCING MS LEE.

THIS WORKBOOK CONTAINS WORKSHEETS, PUZZLE SHEETS AND HOMEWORK ASSIGNMENTS

HI, I'VE BEEN TEACHING MATHS FOR OVER 20 YEARS

I BECAME GOOD AT MATHS THROUGH PRACTICE AND EFFORT

I WILL GUIDE YOU THROUGH THE WORKSHEETS

MATHS IS ABOUT MASTERING A COLLECTION OF SKILLS, AND I CAN HELP YOU DO THIS

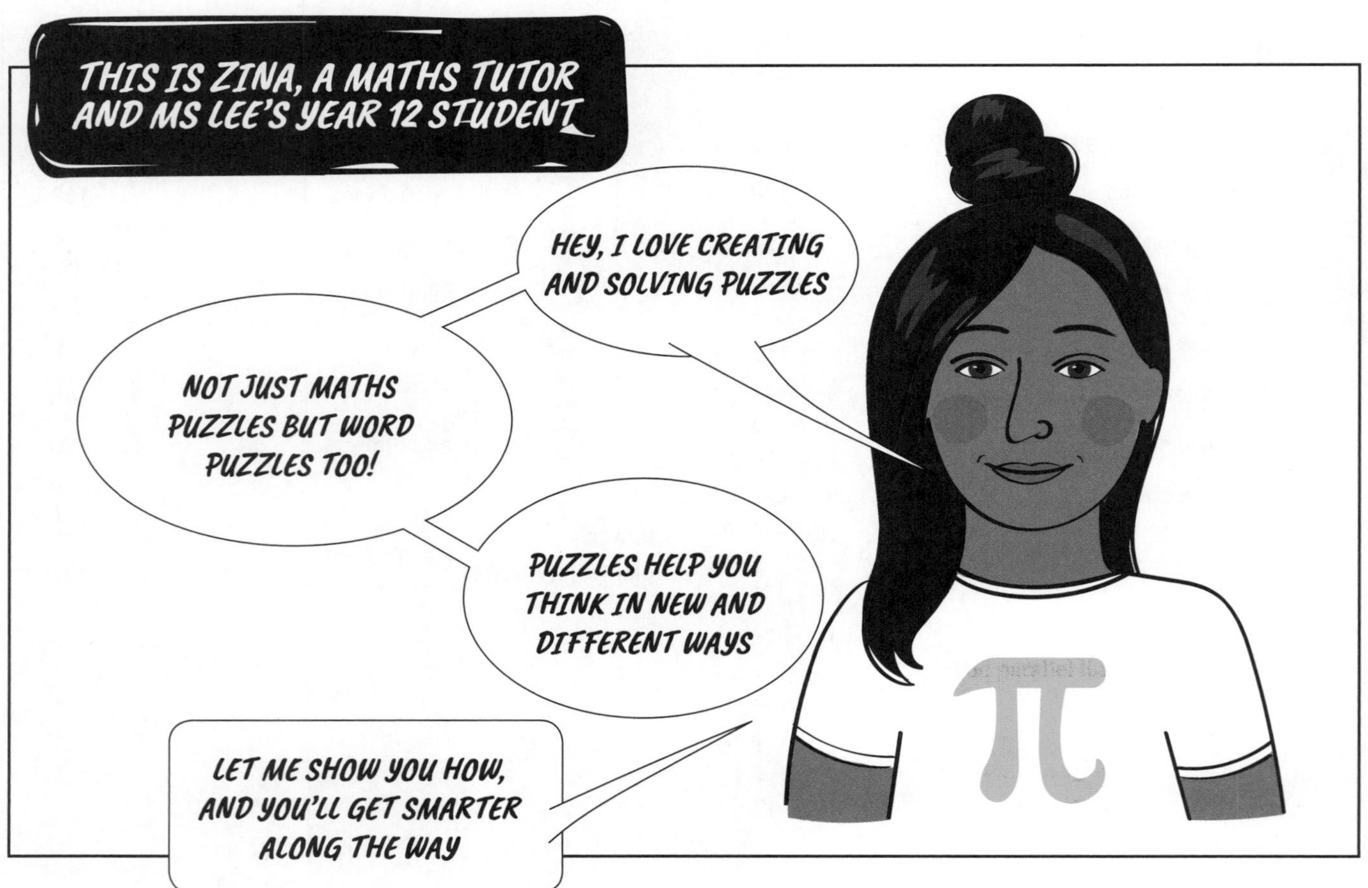

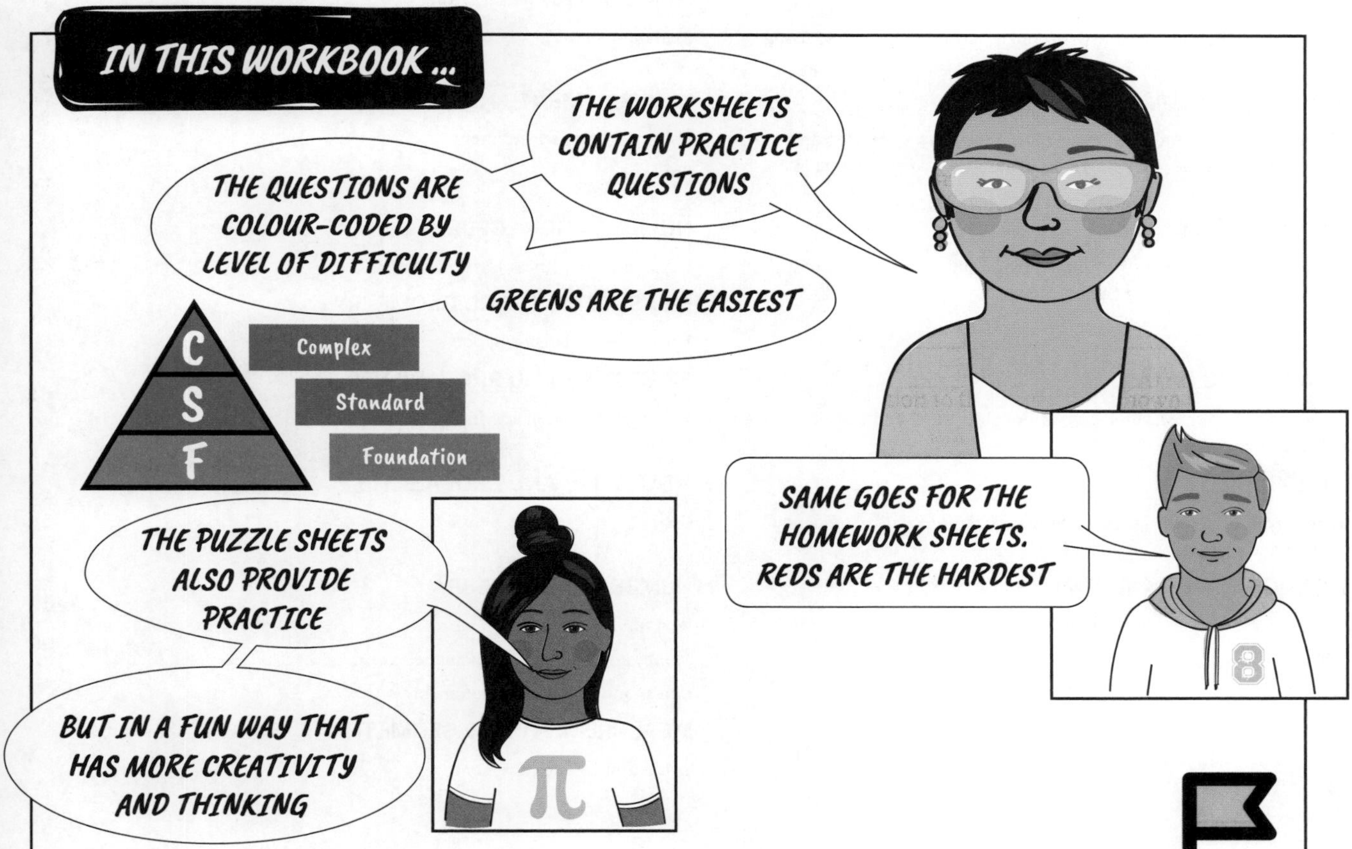

IN THIS WORKBOOK ...

THE WORKSHEETS CONTAIN PRACTICE QUESTIONS

THE QUESTIONS ARE COLOUR-CODED BY LEVEL OF DIFFICULTY

GREENS ARE THE EASIEST

C Complex

S Standard

F Foundation

SAME GOES FOR THE HOMEWORK SHEETS. REDS ARE THE HARDEST

THE PUZZLE SHEETS ALSO PROVIDE PRACTICE

BUT IN A FUN WAY THAT HAS MORE CREATIVITY AND THINKING

SO LET'S GET GOING.

CURRICULUM GRID

Topic skills	Australian curriculum strand and substrand
1 INTEGERS	**NUMBER AND ALGEBRA**
Ordering integers, operations with integers, order of operations	Number and place value
2 ANGLES	**MEASUREMENT AND GEOMETRY**
Classifying angles, complementary and supplementary angles, angles at a point, vertically opposite angles, angles on parallel lines	Geometric reasoning
3 WHOLE NUMBERS	**NUMBER AND ALGEBRA**
Rounding and estimating, multiplying and dividing, divisibility tests. powers and roots, prime and composite numbers, factor trees, HCF and LCM	Number and place value
4 FRACTIONS AND PERCENTAGES	**NUMBER AND ALGEBRA**
Ordering fractions, operations with fractions including mixed numerals, fraction and percentage of a quantity, decimals, percentages, expressing quantities as fractions and percentages	Real numbers
5 ALGEBRA AND EQUATIONS	**NUMBER AND ALGEBRA**
The laws of arithmetic, converting sentences to algebraic expressions, substitution, equations, two-step equations	Number and place value Patterns and algebra Linear and non-linear relationships
6 GEOMETRICAL FIGURES	**MEASUREMENT AND GEOMETRY**
Transformations, line and rotational symmetry, classifying triangles and quadrilaterals, angle sums, exterior angle of a triangle, properties of quadrilaterals	Location and transformation Geometric reasoning
7 DECIMALS	**NUMBER AND ALGEBRA**
Ordering decimals, operations with decimals, terminating and recurring decimals, rounding decimals	Real numbers
8 AREA AND VOLUME	**MEASUREMENT AND GEOMETRY**
Metric system, metric units for area, volume, capacity, perimeter, areas of rectangles, triangles, parallelograms, composite shapes, volume of a rectangular prism	Using units of measurement Shape
9 THE NUMBER PLANE	**NUMBER AND ALGEBRA**
Location using coordinates, the number plane and quadrants, graphing tables of values, transformations on the number plane	Linear and non-linear relationships **MEASUREMENT AND GEOMETRY** Location and transformation
10 ANALYSING DATA	**STATISTICS AND PROBABILITY**
Interpreting graphs, misleading graphs, dot plots, stem-and-leaf plots, mean, mode, median, range	Data representation and interpretation
11 PROBABILITY	**STATISTICS AND PROBABILITY**
Sample spaces, probability, experimental probability, relative frequency, complementary events	Chance
12 RATIOS, RATES AND TIME	**NUMBER AND ALGEBRA**
Ratios, rates, best buys, travel graphs, time calculations, 24-hour time, timetables	Real numbers Money and financial mathematics Linear and non-linear relationships **MEASUREMENT AND GEOMETRY** Using units of measurement

STARTUP ASSIGNMENT 1 (1)

HERE ARE SOME SKILLS YOU NEED TO KNOW TO DO WELL IN MATHS. PART A IS MIXED SKILLS, PART B IS FOR THE INTEGER TOPIC WE'RE STARTING.

WS WORKSHEET

PART A: BASIC SKILLS

/ 15 marks

1 How many sides has a pentagon? ________

2 In 409 675, which digit is in the thousands place? ________

3 Write 409 675 in words.

____

____

____

4 Draw an **obtuse** angle.

5 **a** How many faces has a cube? ________

b What shape is each face? ________

6 Evaluate $10 \times 100 \times 10$. ________

7 List the factors of 27.

____

8 What comes next in this pattern?

25, 21, 17, 13, 9, ________

9 How many cubes are needed to make this figure? ________

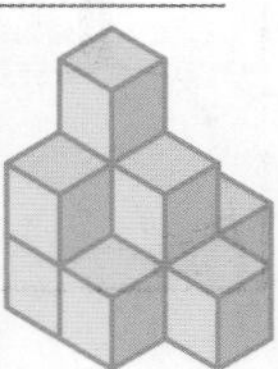

10 How many degrees in a right angle?

____

11 What number plus 75 equals 180?

____

12 Simplify $\frac{16}{24}$. ________

13 Find the average of \$14, \$20 and \$23.

____

14 What does 16:40 in 24-hour time mean?

____

PART B: NUMBER

/ 25 marks

15 Evaluate:

a $16 + 67$ ________

b $25 - 9$ ________

c 4×8 ________

d 7×7 ________

16 Write these numbers in ascending order:

55, 52, 15, 5, 51, 25.

____

17 Write one million in figures.

____

18 What is the difference between 22 and 2?

____

C S F

WS WORKSHEET

19 Evaluate:

a $42 \div 6$ ______

b 9×5 ______

c $73 - 18$ ______

d $1375 \div 5$ ______

20 True or false?

a $4 < 10$ ______

b $3 \times 3 < 8$ ______

c $6 + 9 > 15$ ______

d $15 \div 3 > 5 - 2$ ______

21 Write these numbers in descending order:

7, 707, 77, 87, 770, 78.

______

22 What is the product of 7 and 5? ______

23 Evaluate:

a $12 - 6 \times 2$

______

b $56 \div (4 \times 2)$

______

c $5 \times 2 + 8 \div 2$

______

d $8 \times 3 \times 5$

______

e $\frac{7+5}{7-5}$

______

24 Write a number between 18 and 25 that is:

a prime ______

b divisible by 6 ______

25 What number multiplied by itself equals 16?

______

PART C: CHALLENGE

Bonus / 3 marks

Anh, Benazir, Colin, Dominic and Endip are in the same class.

In a maths test, their marks in descending order were **94**, **88**, **80**, **76** and **65**.

- Anh scored less than Benazir.
- Benazir scored less than Dominic.
- Colin scored more than Dominic.
- Endip scored more than Colin.
- Who scored what mark?

C S F

9780170454452

ADDING AND SUBTRACTING INTEGERS 1

NOW PRACTISE ADDING AND SUBTRACTING INTEGERS. THE NUMBER LINE MIGHT HELP.

WS WORKSHEET

EXAMPLES

$7 + (-4) = 7 - 4 = 3$
$3 - (-2) = 3 + 2 = 5$

−10 −9 −8 −7 −6 −5 −4 −3 −2 −1 0 1 2 3 4 5 6 7 8 9 10

1 $8 - 10 =$ ______ **2** $5 - 9 =$ ______ **3** $2 - 7 =$ ______

4 $3 - 4 =$ ______ **5** $6 - 9 =$ ______ **6** $-6 + 5 =$ ______

7 $-10 + 4 =$ ______ **8** $-3 + 6 =$ ______ **9** $-5 + 8=$ ______

10 $-7 + 1 =$ ______ **11** $-8 - 3 =$ ______ **12** $-1 - 4 =$ ______

13 $-7 - 5 =$ ______ **14** $-10 - 4 =$ ______ **15** $-3 - 5 =$ ______

16 $11 + (-8) =$ ______ **17** $3 + (-3) =$ ______ **18** $8 + (-7) =$ ______

19 $2 + (-6) =$ ______ **20** $10 + (-1) =$ ______ **21** $-8 + 7 =$ ______

22 $-2 + 5 =$ ______ **23** $-9 + 10 =$ ______ **24** $-9 + 9 =$ ______

25 $-4 + 7 =$ ______ **26** $5 + (-12) =$ ______ **27** $6 + (-9) =$ ______

28 $11 + (-10) =$ ______ **29** $-4 + (-4) =$ ______ **30** $0 + (-5) =$ ______

31 $4 - 8 =$ ______ **32** $10 - 13 =$ ______ **33** $-5 - 9=$ ______

34 $7 - 11 =$ ______ **35** $1 - 3 =$ ______ **36** $-7 + (-10) =$ ______

37 $-1 + (-5) =$ ______ **38** $-9 + (9) =$ ______ **39** $-3 + (-8) =$ ______

40 $-4 + (-6) =$ ______ **41** $7 - (-1)=$ ______ **42** $6 - (-9) =$ ______

43 $5 - (-5) =$ ______ **44** $2 - (-3) =$ ______ **45** $6 - (-2) =$ ______

46 $-1 - (-4) =$ ______ **47** $-8 - (-7) =$ ______ **48** $-3 - (-5) =$ ______

49 $-3 + (-5) =$ ______ **50** $-7 + 4 =$ ______ **51** $0 + (-3) =$ ______

52 $-2 + 7 =$ ______ **53** $-6 - 4 =$ ______ **54** $2 - (-3) =$ ______

55 $6 + (-4) =$ ______ **56** $-1 + (-5) =$ ______ **57** $-5 - (-4) =$ ______

58 $-1 + 10 =$ ______ **59** $-3 - (-2) =$ ______ **60** $-5 + (-4) =$ ______

61 $7 + (-3) =$ ______ **62** $7 - 12 =$ ______ **63** $-10 - 1 =$ ______

64 $-2 - 7 =$ ______ **65** $-9 + 2 =$ ______ **66** $-5 + (-10) =$ ______

67 $3 - (-3) =$ ______ **68** $-7 + 7 =$ ______ **69** $1 - 5 =$ ______

70 $-1 + (-7)=$ ______ **71** $-2 + 8 =$ ______ **72** $-6 + (-6) =$ ______

1 INTEGER REVIEW

THIS REVISION SHEET COVERS MOST OF THE INTEGERS TOPIC. ASK YOUR TEACHER OR ANOTHER STUDENT IF YOU GET STUCK.

WS WORKSHEET

1 Write these integers in descending order:

$-6, 0, 4, -3, 5, 6, -8, -1$

____

2 Evaluate each expression.

a $3 + (-5)$ ______

b $-3 + 5$ ______

c $3 - 5$ ______

d -3×5 ______

e $-3 - 5$ ______

f $3 - (-5)$ ______

g $-3 + (-5)$ ______

h $-3 \times (-5)$ ______

i $-3 - (-5)$ ______

3 Find the temperature when:

a 8°C drops by 10° ______

b −5°C rises by 4° ______

c −9°C drops by 2° ______

d −1°C rises by 7° ______

e 6°C drops by 9° ______

f −10°C rises by 3° ______

4 Write these integers in ascending order:

$-15, 11, -2, 2, 8, 20, -7, -10$

____

5 Evaluate each expression.

a $18 \div (-2) \times (-3)$ ______

b $\frac{45}{-9}$ ______

c $-4 \times (-7) \times (-2)$ ______

d $7 - 10 - 4$ ______

e $-4 \times (-4) + 3 \div (-1)$ ______

f $-2 + 2 - 8 + 8$ ______

g $-3 + 9 + (-2) - 5$ ______

h $\frac{3 \times (-10)}{-6}$ ______

i $-5 \times [7 - (-3 + 4)]$ ______

j $4 \times (-2) + 6$ ______

k $28 - 8 \times 3$ ______

l $-11 + 5 \times (-3) \div 3$ ______

6 Complete each table.

a

+	−2	7	0	−9	−3
1					
−5					
8					
3					
−1					

b

×	3	−6	2	10	−7
5					
−4					
−2					
7					
−1					

C S F

9780170454452

7 Find the difference when the temperature changes from:

a 2°C to 14°C ____________

b −3°C to 5°C ____________

c 6°C to −1°C ____________

d −4°C to −8°C ____________

e −5°C to 0°C ____________

f −9°C to −1°C ____________

8 Evaluate each expression.

a $-3 \times 6 + (-4) \times 6$ ____________

b $-7 - (4 - 9) - 10$ ____________

c $5 \times (-2) + 4 \times (-1) + 12$ ____________

d $2 \times (-4) \times 8$ ____________

e $\frac{-24+4}{4-8}$ ____________

f $(-4)^2 + (-3)^2$ ____________

g $(-3 - 3) \times 10$ ____________

h $14 - 18 \div (-2)$ ____________

i $[3 \times (7 - 10) + 5] \div 4$ ____________

j $12 - [-8 \div (-1) + 2]$ ____________

k $18 + 3 - 4 \times 7$ ____________

l $-5 \times 8 - 27 \div (-3)$ ____________

WS WORKSHEET

C S F

1 INTEGERS FIND-A-WORD

CAN YOU SEE THE WORDS FROM THE INTEGERS TOPIC HIDDEN IN THE PUZZLE?

I	V	G	N	I	D	N	E	C	S	E	D	Y	N	X	R	G	E	S	L	A	B	I	C	K
I	H	R	A	I	Q	H	M	Q	D	D	O	R	E	Z	H	J	L	E	A	I	N	P	J	R
Y	O	U	B	E	V	I	T	A	G	E	N	F	G	N	I	D	N	E	C	S	A	Y	S	U
H	J	Q	S	J	K	T	K	V	L	E	A	D	J	E	V	I	T	I	E	L	O	H	W	M
V	N	I	W	O	H	B	A	K	W	S	U	N	I	M	Q	U	A	D	R	A	N	T	U	D
Q	G	S	A	W	E	C	N	E	R	E	F	F	I	D	D	F	D	Z	O	A	S	I	O	E
I	R	O	R	D	E	R	E	D	C	P	K	T	R	E	B	M	U	N	M	V	E	X	T	R
M	D	O	D	E	T	C	E	R	I	D	T	D	F	G	R	E	A	T	E	R	B	A	S	N
S	Q	L	H	B	A	D	U	V	X	G	X	S	E	T	A	N	I	D	R	O	U	C	Y	J
P	D	V	T	T	E	N	M	D	M	M	S	I	X	A	O	R	I	G	I	L	S	M	B	T
J	L	I	I	W	W	G	E	Y	S	S	S	E	L	E	M	M	R	N	A	P	E	M	D	C
Q	V	A	W	K	J	P	O	S	I	T	I	V	E	S	H	N	N	V	U	A	E	V	J	U
K	S	Y	N	G	O	O	R	T	G	C	W	P	R	W	O	W	E	N	I	L	R	A	C	D
W	S	E	L	S	J	P	L	U	S	V	K	G	I	I	T	O	I	X	Q	T	G	C	L	O
X	N	L	I	I	P	R	V	L	T	H	R	I	T	P	O	Z	B	S	T	Z	E	O	I	R
K	X	T	F	O	R	E	G	E	T	N	I	A	I	B	Q	H	U	V	G	Q	D	G	B	P
E	S	E	X	A	Y	O	E	S	C	Z	R	Q	U	O	T	I	E	N	T	N	I	O	P	C
C	Q	J	C	M	M	A	G	N	I	E	U	D	E	F	C	A	N	G	N	X	C	D	Z	H
D	P	F	K	U	M	F	C	T	P	T	X	X	X	D	I	X	W	I	D	S	Y	A	N	C
N	E	T	I	S	O	P	P	O	Z	M	R	W	J	N	P	Y	T	S	U	I	S	L	E	C

Find these words in the puzzle above. They are across, up and down, and diagonal, and can be backwards as well as forwards.

ASCENDING	CELSIUS	DEGREES	DEPOSIT
DESCENDING	DIFFERENCE	EVALUATE	GREATER
INTEGER	LESS	LINE	MINUS
MORE	NEGATIVE	NUMBER	OPERATION
OPPOSITE	ORDER	PLUS	POSITIVE
PRODUCT	QUOTIENT	SIGN	SUM
WHOLE	WITHDRAW	ZERO	

WHAT IS THE QUESTION? 1

THIS WORKSHEET IS BACK-TO-FRONT. WE GIVE THE ANSWER, YOU GIVE THE QUESTION.

WS WORKSHEET

Each number below is an **answer** to a question involving 2 numbers. For each answer, write 4 questions: one each involving addition (+), subtraction (−), multiplication (×) and division (÷).

EXAMPLE

The answer is −2.

$-5 + 3 = -2$ $\quad 5 - 7 = -2$ $\quad -1 \times 2 = -2$ $\quad -8 \div 4 = -2$

−3	−1	0
−5	−10	6
−9	−20	−75
$-\frac{1}{2}$	−0.75	2.5

CHALLENGE

For each answer, find 3 questions involving 3 numbers and 2 operations (+, −, ×, ÷).

−1	−6	−12

1 INTEGERS 1

THIS IS YOUR WEEKLY HOMEWORK ASSIGNMENT, COVERING THE CURRENT TOPIC (INTEGERS) AS WELL AS MIXED REVISION. NO CALCULATORS IN PART A!

Name:

Due date:

Parent's signature:

Part A	/ 8 marks
Part B	/ 8 marks
Part C	/ 8 marks
Part D	/ 8 marks
Total	/ 32 marks

HW HOMEWORK

PART A: MENTAL MATHS

Calculators are not allowed

1 Evaluate $95 \div 5$.

____

2 How many hours in one week? __________

3 Describe an acute angle in words.

____

____

4 A square has sides that are 6 cm long.
What is its area?

____

5 Find the perimeter of this shape.

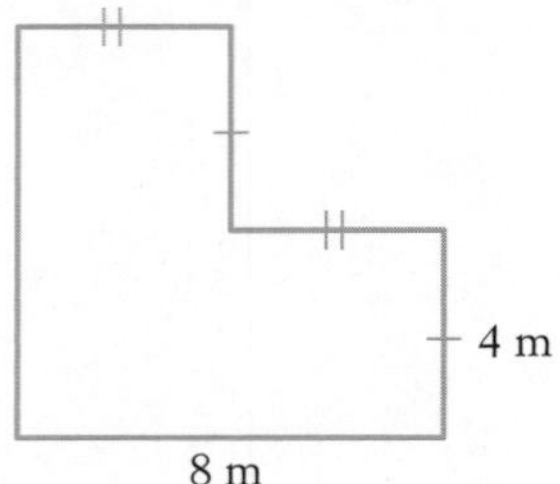

____

6 What is the general name for a shape with 4 sides?

____

7 Write down all the multiples of 7 between 20 and 40.

____

8 How many degrees in a straight angle?

____

PART B: REVIEW

1 Evaluate each expression.

a $28 - 5 \times 5$

____

b $(28 - 5) \times 5$

____

c $9 + 4 - 8 - 1$

____

d $9 + 4 - (8 - 1)$

____

2 Complete:

a ______ $-9 = 14$

b $14 -$ ______ $= 9$

3 Complete each blank with a $<$ or $>$ sign.

a 15 ______ 18

b $4 + 7$ ______ $9 - 2$

C
S
F

9780170454452

PART C: PRACTICE

› Integers on a number line
› Ordering integers
› Adding integers

1 (2 marks) Write the positions of -4, -1, 3 and 5 on this number line.

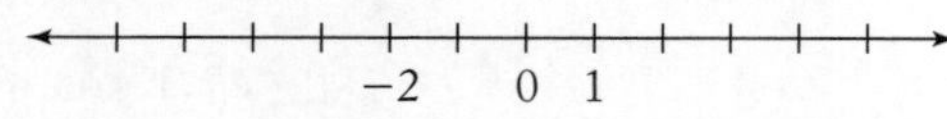

2 Complete each blank with a < or > sign.

a -4 ______ 3

b 5 ______ -1

c -2 ______ 0

3 Evaluate each sum.

a $-3 + 4$

______

b $5 + (-8)$

______

c $-9 + 6$

______

PART D: NUMERACY AND LITERACY

1 a Write these integers in descending order:

$-2, 8, -6, -3, 4$

______

b Find the sum of the above integers.

______

2 Complete: On a number line, the numbers on the right are ______ than the numbers on the left.

3 Write in words the opposite of each statement.

a Winning $50

______

b 8 degrees above zero

______

4 Taylor took 4 steps forward, then 1 step backwards. She did this 4 times. What is her position now compared to where she started?

______

5 Which integer's opposite is itself? ______

6 Why isn't -3.5 an integer?

______

______

HW HOMEWORK

C
S
F

9780170454452

1 INTEGERS 2

JUST LIKE IN SPORT, DRILL AND PRACTICE IS IMPORTANT IN MATHS. PART B OF THIS HOMEWORK ASSIGNMENT IS REVIEW OF LAST WEEK'S WORK, PART C IS PRACTICE.

Name:

Due date:

Parent's signature:

Part A	/ 8 marks
Part B	/ 8 marks
Part C	/ 8 marks
Part D	/ 8 marks
Total	/ 32 marks

PART A: MENTAL MATHS

Calculators are not allowed.

1 Evaluate $8 + 6 \times 2$.

2 Complete: 320 cm = ______ m

3 Describe an obtuse angle in words.

4 A concert began at 7:50 p.m. and went for 2 hours 15 minutes. At what time did it finish?

5 A square has a perimeter of 100 cm. What is the length of each side?

6 Evaluate $351 \div 9$.

7 **a** Draw a rectangular prism.

b How many faces has a rectangular prism?

PART B: REVIEW

1 (2 marks) Complete this number line.

-4

2 Complete each blank with a $<$ or $>$ sign.

a -2 ______ -4

b -1 ______ 3

3 Write these integers in descending order:

$-5, 2, 17, 0, -2, -8$.

4 Evaluate each sum.

a $-12 + 5$

b $-9 + (-4)$

5 Complete: ______ $+ 2 = -10$

C S F

9780170454452

PART C: PRACTICE

› Subtracting integers
› Integers revision

1 Evaluate each difference.

a $6 - 8$

b $-3 - 4$

c $9 - (-3)$

2 **a** Write these integers in ascending order:

$-3, 4, 2, -1, 0$

b Find the sum of the above integers.

3 Evaluate each sum.

a $-1 + (-7)$

b $-6 + 10$

4 Write in words the opposite of '30 metres above sea level'.

C
S
F

PART D: NUMERACY AND LITERACY

1 The temperature was $-3°$ C this morning but it went up 7 degrees. What is the new temperature?

2 Josh was in a lift at a big city department store. He started at ground level, went up 3 levels for toys, down 2 levels for men's fashions, then up 4 levels to the cafeteria for lunch.

a On which level was the cafeteria?

b After lunch, Josh went down 7 levels to the the car park. How far below ground level is the car park?

3 Write 2 positive and 2 negative integers that altogether have a sum of -5.

4 The temperatures for 5 cities for one day in January are shown:

Bouganville	31°C
Melbourne	29°C
Moscow	−7°C
New York	−1°C
Paris	−2°C

a Which was the second coldest city?

b How much hotter was it in Melbourne than New York?

c How much colder was it in Moscow than in New York?

d Find the average of the 5 temperatures.

WS WORKSHEET

2 STARTUP ASSIGNMENT 2

THIS ASSIGNMENT COVERS BASIC SKILLS AND THINGS YOU SHOULD KNOW BEFORE LEARNING THE ANGLES TOPIC.

PART A: BASIC SKILLS

/ 15 marks

1 Evaluate:

a 13×30 ______

b $\frac{4}{5} \times 40$ ______

2 How many axes of symmetry does a square have? ______

3 Is 359 divisible by 3? ______

4 Complete: 1.62 m = ______ mm

5 What is another name for 100 years?

______

6 What comes next in this pattern?

4, 8, 16, 32, 64, ______

7 Draw a cylinder.

8 List the factors of 36.

______

9 Is 8.17 closer to 8.1 or 8.2? ______

10 How many vertices (corners) has a rectangular prism? ______

11 Calculate the area of this triangle.

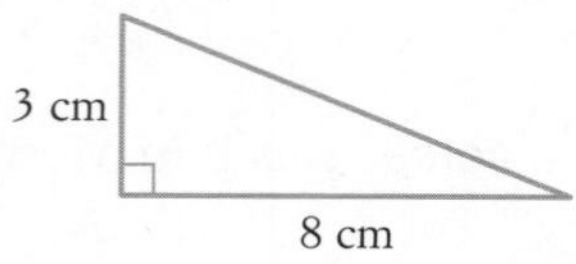

______

12 Write in figures 'Thirteen million, nine hundred thousand, five hundred and twenty-eight'.

______

13 What does the prefix 'kilo' mean?

______

14 What is the time 3 hours and 10 minutes after 5:30 p.m.?

______

PART B: ANGLES AND SHAPES

/ 25 marks

15 What is the opposite of north-east?

______

16 How many degrees in a:

a straight angle? ______

b three-quarter turn? ______

c revolution? ______

17 Draw a reflex angle.

18 Evaluate:

a $180° - 72°$ ______

b $180° \div 4$ ______

c $90° - 38°$ ______

d $360° - 145° - 95°$ ______

C S F

9780170454452

19 Measure the size of the marked angle.

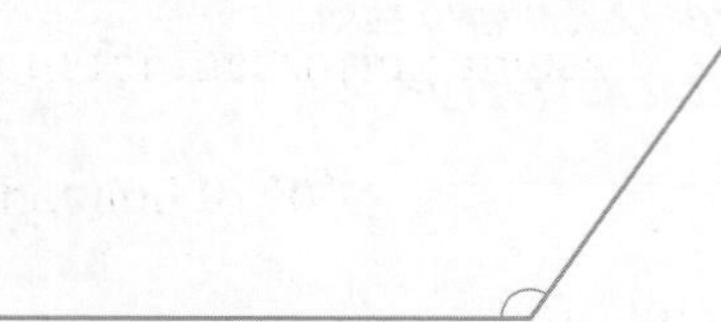

20 What is another name for a 90° angle?

---

21 Is this angle obtuse or acute?

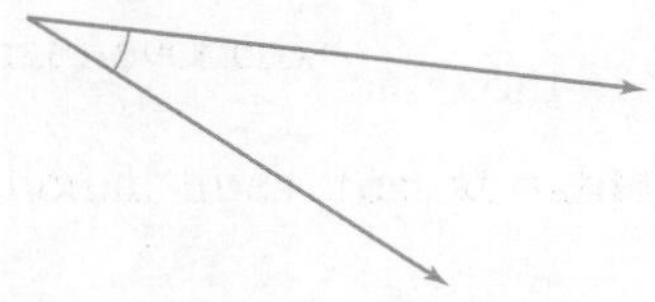

---

22 Draw a 180° angle.

23 Which angle is the largest: *A*, *B* or *C*?

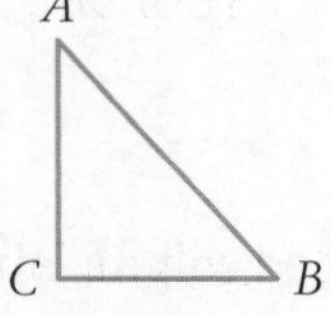

---

24 **a** What is the name of this shape?

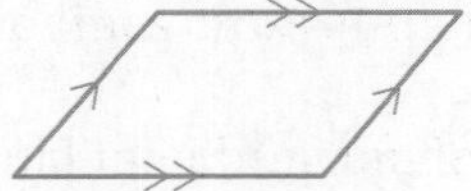

---

b What do the arrows on the sides of the shape mean?

---

25 How many angles has an octagon? ______

26 **a** What type of triangle has 3 equal sides and 3 equal angles? ______

b If the sum of the angles in the triangle from part **a** is 180°, what is the size of each angle?

---

27 Use a protractor to construct a 49° angle.

28 What type of angle measures between 180° and 360°?

---

29 True or false? 'A straight angle is larger than an obtuse angle.' ______

30 In this diagram, which angle matches angle *D* (angle *E*, *F*, *G* or *H*)?

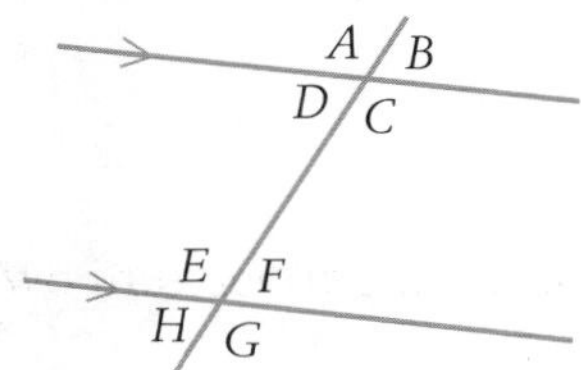

---

31 **a** What type of shape has 4 equal sides and 4 equal angles?

---

b What is the size of one of the angles in the shape from part **a**?

PART C: CHALLENGE

Bonus / 3 marks

A square has 2 diagonals.

A pentagon has 5 diagonals.
How many diagonals has a 20-sided shape?

(*Hint*: Look for a pattern.)

C S F

WS WORKSHEET

2 ESTIMATING AND MEASURING ANGLES

YOU NEED A PROTRACTOR, RULER AND EYES FOR THIS PRACTICAL ACTIVITY.

Follow these steps for each angle below.

1. Draw an angle of that size by estimating (without using a protractor).
2. When directed by your teacher, use a protractor to measure the angle and write the answer in the table on the next page.
3. Calculate the error (how many degrees you were off by) and write it in the table.

1 60°	**2** 45°	**3** 80°
4 58	**5** 160°	**6** 89°
7 100°	**8** 123°	**9** 175°

Question	1	2	3	4	5	6	7	8	9	
Angle	60°	45°	80°	5°	160°	89°	100°	123°	175°	
Size of angle drawn										
Error										Total error

2 FIND THE UNKNOWN ANGLE

Find the value of the variable(s) in each diagram, giving reasons.

1

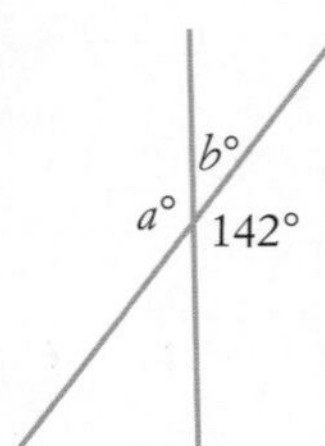

2

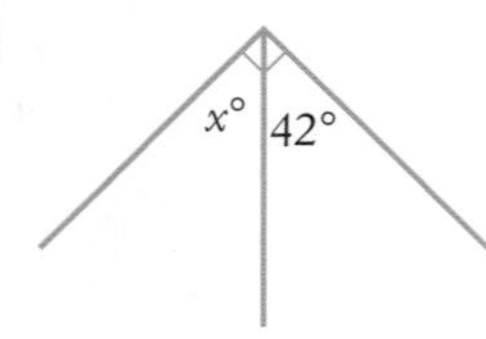

3

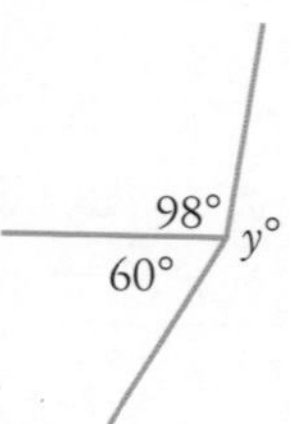

4

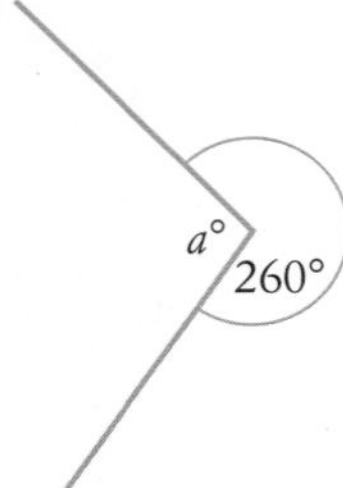

5

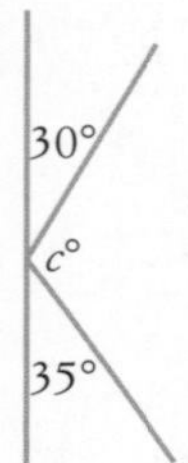

6

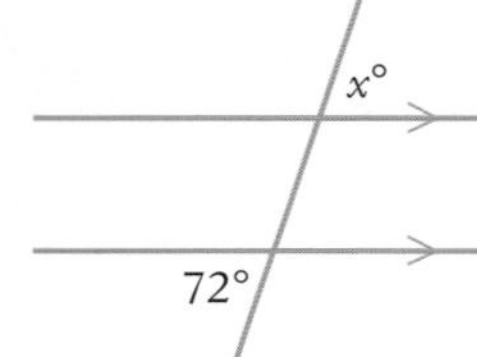

7

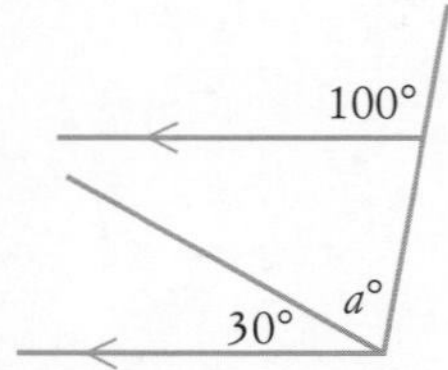

8

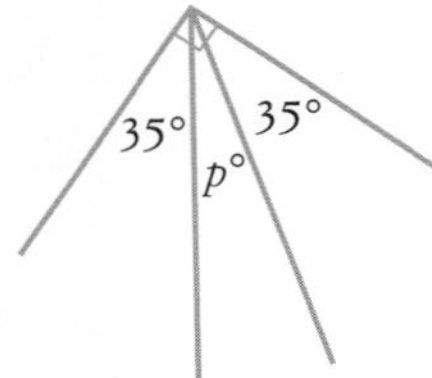

9

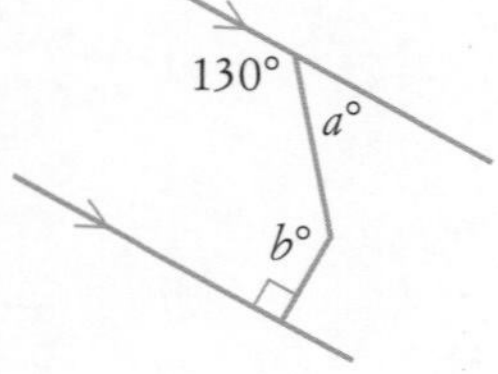

10

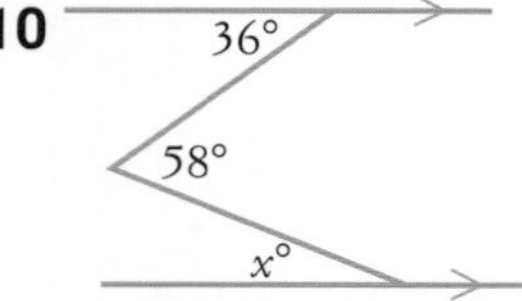

 9780170454452

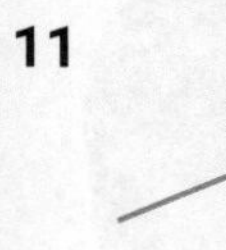

11

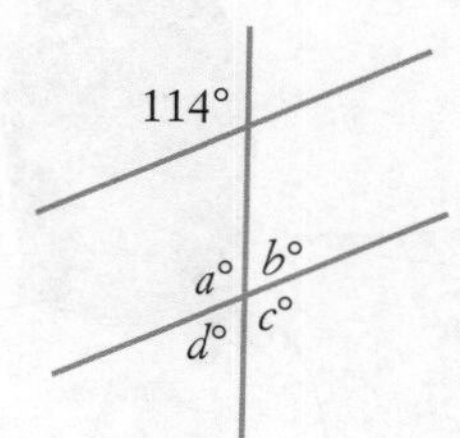

12

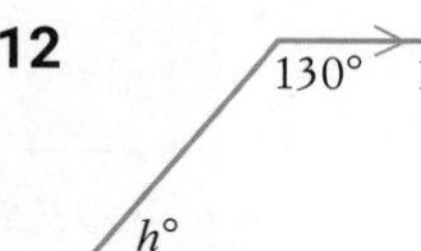

13

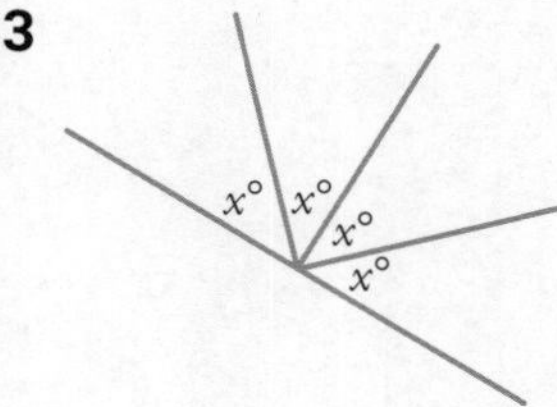

14

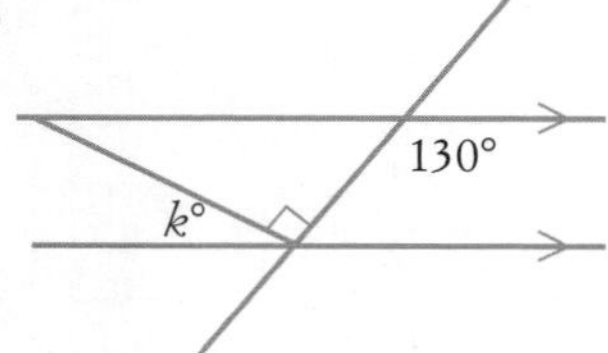

15

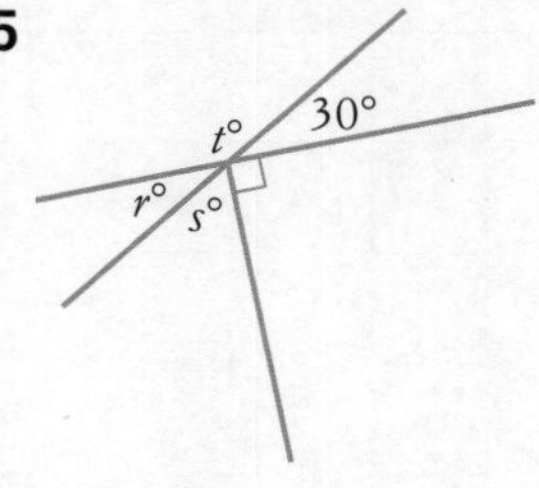

16

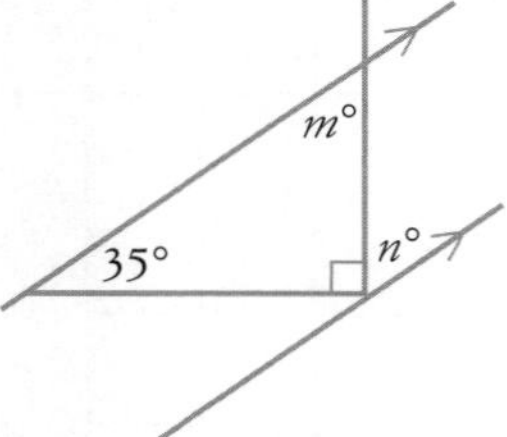

17

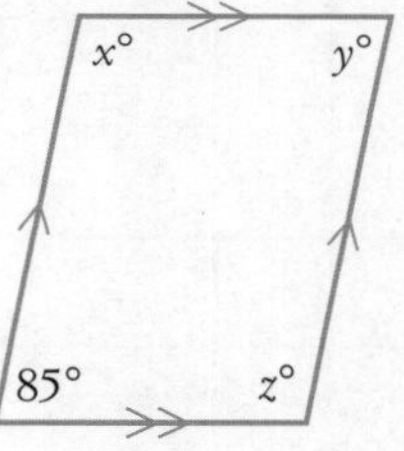

18

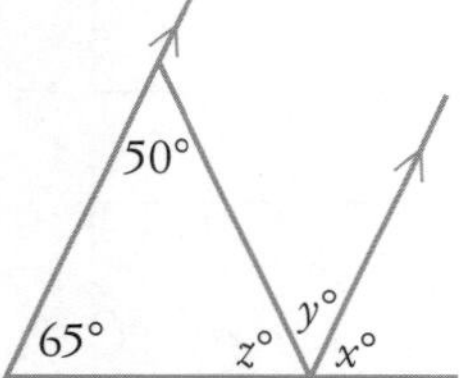

19

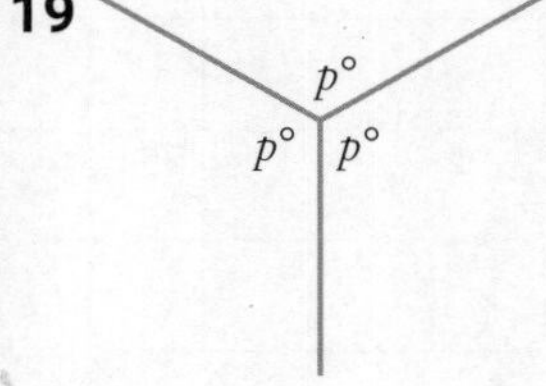

20

x° x°

100°

2 ANGLES CROSSWORD

DO YOU REMEMBER THE TERMINOLOGY USED IN THIS TOPIC? THIS CROSSWORD WILL HELP, BUT SOME OF THE CLUES ARE TRICKY.

1 2 3 4 5 6 7 8 9 10 11 12 13 14 15 16 17 18 19 20 21 22 23 24 25 26 27

Across

3 Instrument for measuring angles

8 Angles that add to 90°

10 A line that crosses 2 or more other lines

12 The 'corner' of an angle or shape

13 The number of angles created when 2 lines cross

16 Lines that never cross

17 Perpendicular lines cross at right ______.

18 The measuring unit for angle size

20 A right angle is a quarter- ______.

22 180° angle

24 Angle between 180° and 360°

25 How many straight angles in a revolution?

26 A protractor has a clockwise and an anticlockwise s ______.

27 The number of degrees in a right angle

PS PUZZLE SHEET

Down

1 A part of an angle that is also a part of your body

2 Angles that add to 180°

4 360° angle

5 These 'inside' angles between parallel lines cut by a transversal are supplementary.

6 Lines that intersect at 90°

7 Two lines cross at their point of i ______.

8 Equal matching angles on parallel lines cut by a transversal

9 Angles between parallel lines on opposite sides of a transversal.

11 Two intersecting lines create 2 pairs of ______ opposite angles.

12 Angle between 90° and 180°

15 Neighbouring angles that share an arm

17 Smaller than a right angle

19 Vertically opposite angles are ______.

21 The angle a clock's minute hand turns in 10 minutes

23 This curved line is used to mark an angle.

2 ANGLES

NOW LET'S PRACTISE AND REVISE OUR ANGLE SKILLS. WHAT'S AN ANGLE THAT IS LESS THAN 90° CALLED AGAIN?

Name:

Due date:

Parent's signature:

Part A	/ 8 marks
Part B	/ 8 marks
Part C	/ 8 marks
Part D	/ 8 marks
Total	/ 32 marks

PART A: MENTAL MATHS

Calculators not allowed.

1 Evaluate 3.21 + 25.6.

2 Find the perimeter of this rectangle.

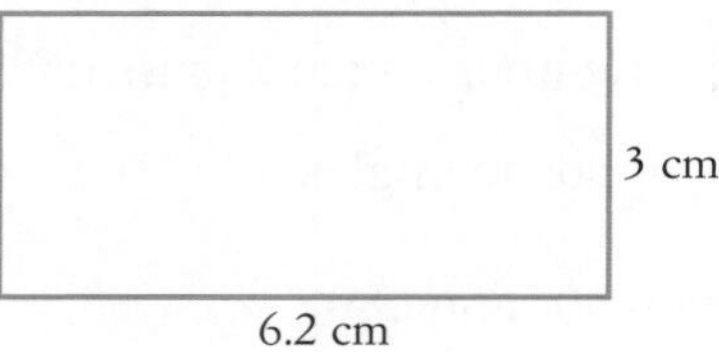

3 How many axes of symmetry does a rectangle have?

4 How many months of the year begin with the letter A?

5 Evaluate 12 + 20 ÷ 4.

6 What is the time 3 hours 20 minutes after 8:15 a.m.?

7 Evaluate 6 + (−3).

8 Complete: ______ − 12 = −5.

PART B: REVIEW

1 Draw and mark a 90° angle.

2 What is the special name for a 90° angle?

3 How many degrees in:

a a straight angle? ______

b a full revolution? ______

4 What is an acute angle?

5 Name the instrument used for drawing and measuring angles.

C S F

HW HOMEWORK

9780170454452

6 Measure this angle.

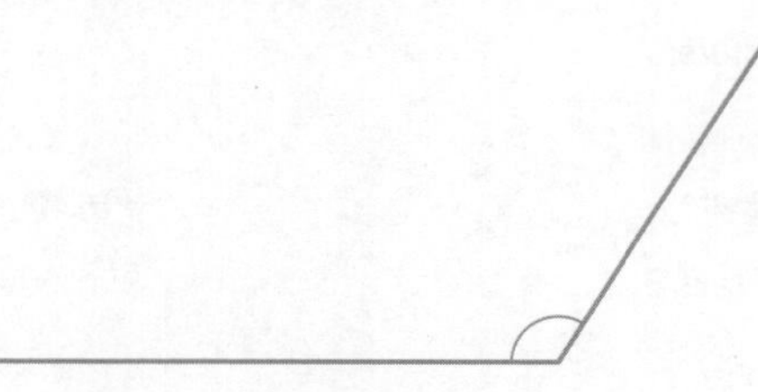

7 Construct a 35° angle.

PART C: PRACTICE

› Naming angles
› Measuring and drawing angles
› Classifying angles

1 (2 marks) Draw an obtuse angle and label it $\angle ZVR$.

2 **a** Name the marked angle below. ______

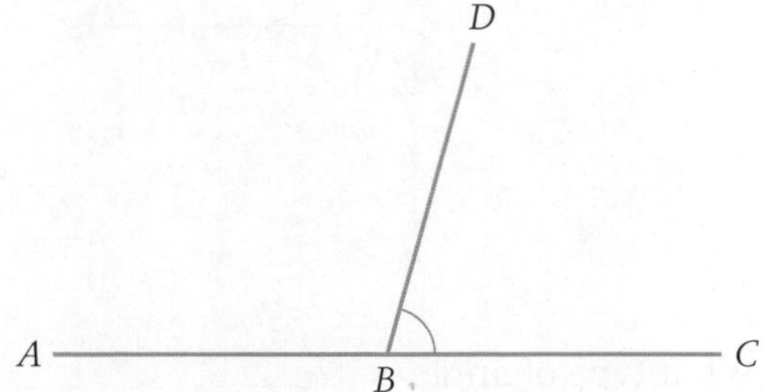

b Name its adjacent angle. ______

c Measure the marked angle. ______

d Name the vertex of the angle. ______

3 Construct a 125° angle.

4 What is a reflex angle?

______

______

PART D: NUMERACY AND LITERACY

1 The angle between the hands of a clock has been marked.

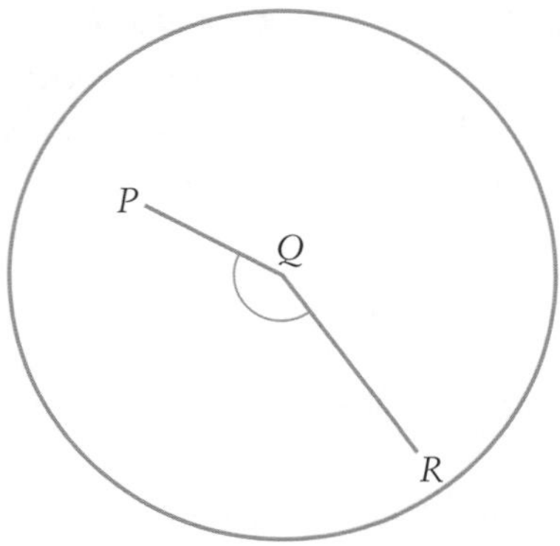

a What type of angle is marked?

______

b (2 marks) Complete: QP and QR are called the ______ of the angle while Q is called the ______.

c Measure the size of the angle. ______

d How much bigger is it than a right angle?

______.

e How much smaller is it than a straight angle?

______.

2 Write 2 angle sizes whose sum is:

a an obtuse angle

______

b an acute angle

______

HW HOMEWORK

2 ANGLE GEOMETRY

PART D QUESTIONS CAN BE COMPLEX BECAUSE THEY ASK YOU TO WRITE ABOUT YOUR MATHS, USING THE RIGHT TERMINOLOGY. DO YOU KNOW THAT 2 ADJACENT RIGHT ANGLES FORM A STRAIGHT ANGLE?

Name:

Due date:

Parent's signature:

Part A	/ 8 marks
Part B	/ 8 marks
Part C	/ 8 marks
Part D	/ 8 marks
Total	/ 32 marks

PART A: MENTAL MATHS

Calculators not allowed.

1 **a** Draw an isosceles triangle.

b How many axes of symmetry does an isosceles triangle have?

____

2 Write '10 minutes to 3 in the afternoon' in 24-hour time.

____

3 Evaluate 32 × 6.

____

4 Evaluate 80 − 8 × 9.

____

5 Name this solid.

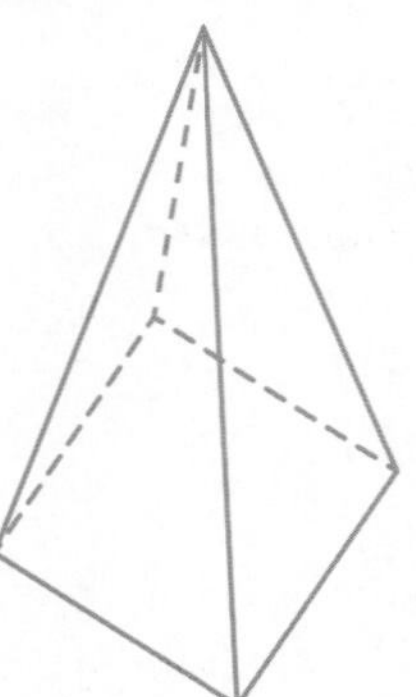

____

6 Find 20% of $45.

____

7 Shade 0.4 of this rectangle.

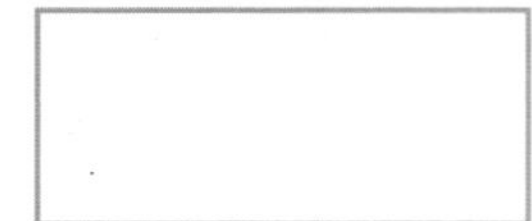

PART B: REVIEW

1 **a** What angle is marked? ______________________

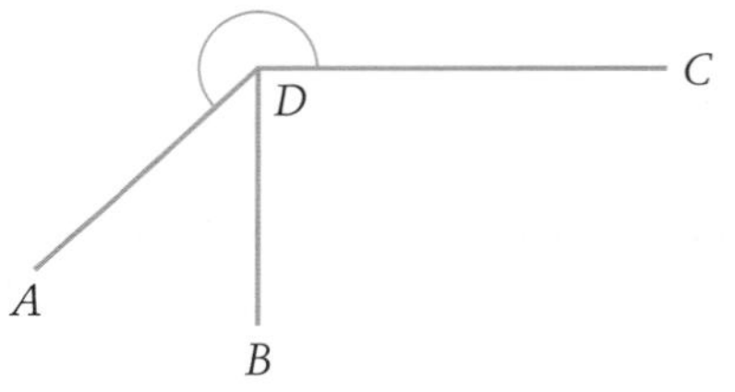

b What type of angle is it? ______________________

c Estimate its size. ______________________

2 Draw a 360° angle.

3 What is the name for:

a an angle between 90° and 180°?

____

b a 180° angle?

____

9780170454452

4 (2 marks) Construct a 145° angle and label it $\angle KTM$.

PART C: PRACTICE

- Complementary and supplementary angles
- Angles at a point
- Vertically opposite angles
- Constructing parallel lines

1

For this rhombus, name a pair of:

a parallel intervals ______________________

b perpendicular intervals ________________

2 Find the value of each variable.

a

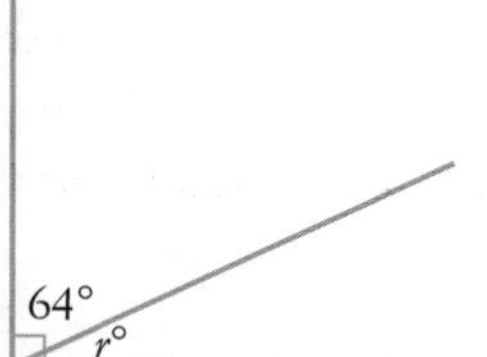

b

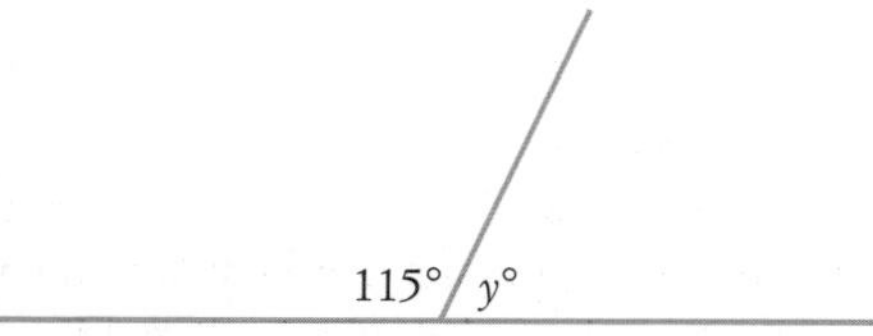

c

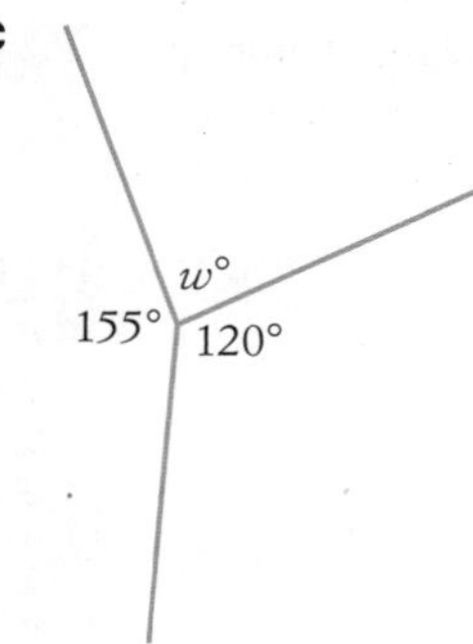

d

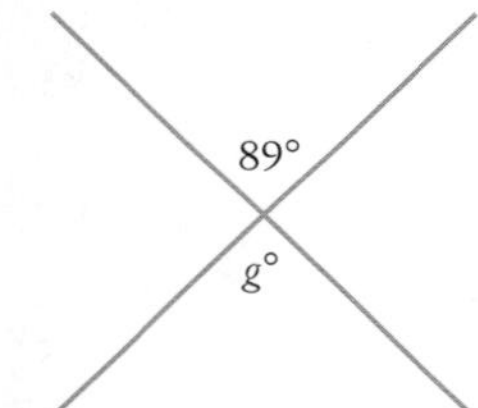

3 Draw a line through X parallel to the line.

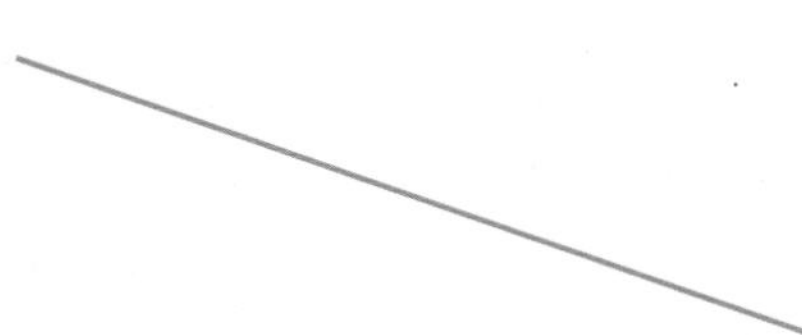

4 Write the symbol for 'is parallel to'. ____________

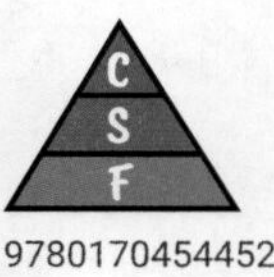

PART D: NUMERACY AND LITERACY

1 Draw a pair of parallel lines.

2 Describe where in everyday life you might see parallel lines.

3 (2 marks) Find the value of x and y, giving reasons.

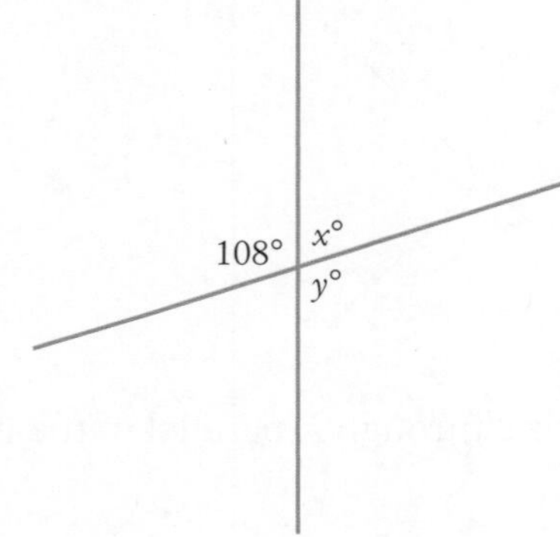

4 Complete: Two angles are complementary if

5 What type of angle is related to the word **perpendicular**?

6 What are perpendicular lines?

7 Write the algebraic rule that relates a, b and c regarding the sum of the angles at a point.

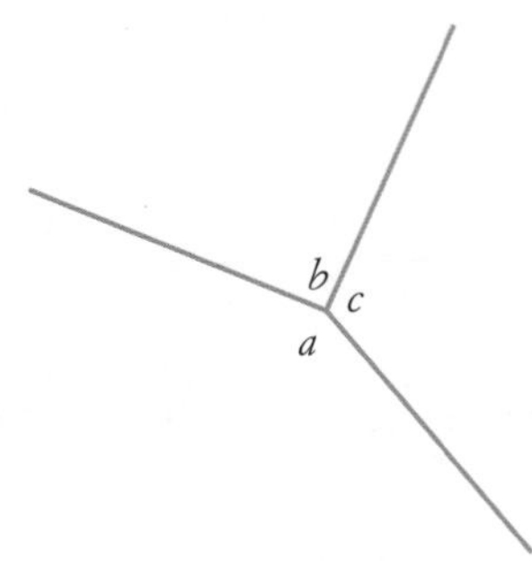

HW HOMEWORK

C
S
F

9780170454452

Name:
Due date:
Parent's signature:

Part A	/ 8 marks
Part B	/ 8 marks
Part C	/ 8 marks
Part D	/ 8 marks
Total	/ 32 marks

ANGLES ON PARALLEL LINES 2

PART A: MENTAL MATHS

Calculators not allowed.

1 Write the next number in this pattern: 8, 4, 2, 1,

______

2 Evaluate 9×12.

______

3 What is an octagon?

______

4 A plane flight started at 8:05 p.m. and lasted 2 hours 15 minutes. What time did it finish?

______

5 Complete: 685 cm = ______ m.

6 Evaluate $3 \times (-7)$. ______

7 Find the area of this rectangle.

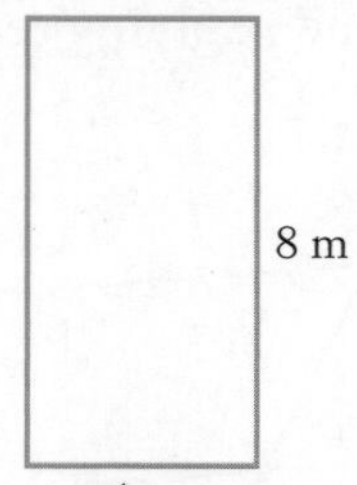

______

8 What is the chance of rolling an odd number on a playing die?

______

C S F

PART B: REVIEW

1 What angle size is complementary to 75°?

______

2 Find the value of each variable.

a

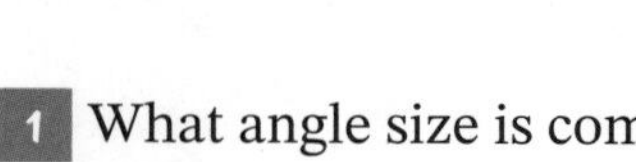

______

b

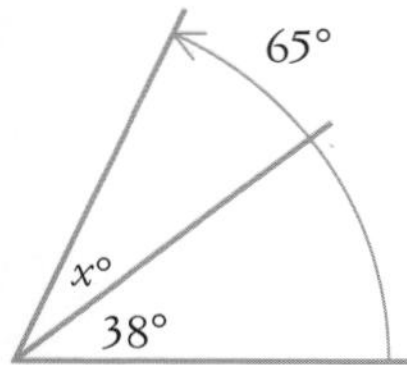

______

c

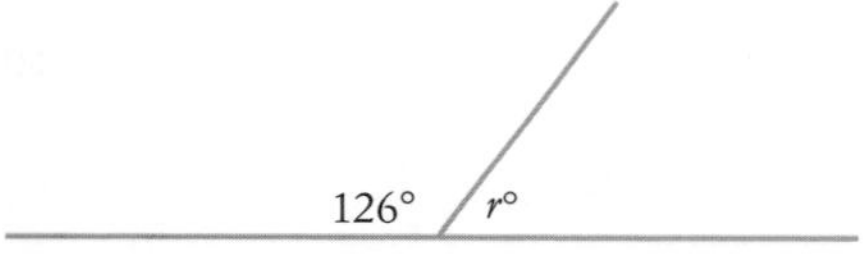

______

3 **a** Draw a pair of vertically opposite angles.

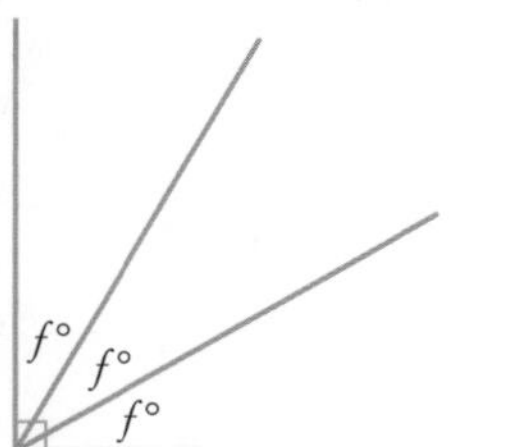

b Write the geometrical rule about vertically opposite angles.

______

______

HW HOMEWORK

4 **a** Draw a reflex angle.

b If a reflex angle is 211°, what is the size of its adjacent angle?

____

PART C: PRACTICE

› Corresponding, alternate and cointerior angles on parallel lines

1 **a** What is the special name of the interval AB in the diagram?

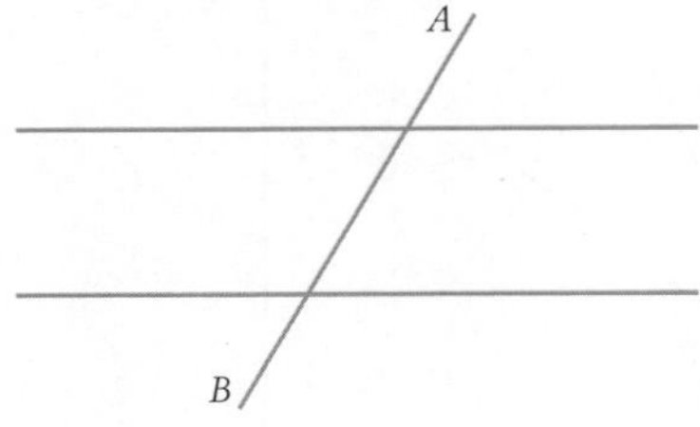

____

b Mark a pair of alternate angles in the diagram.

2 Find the value of each variable.

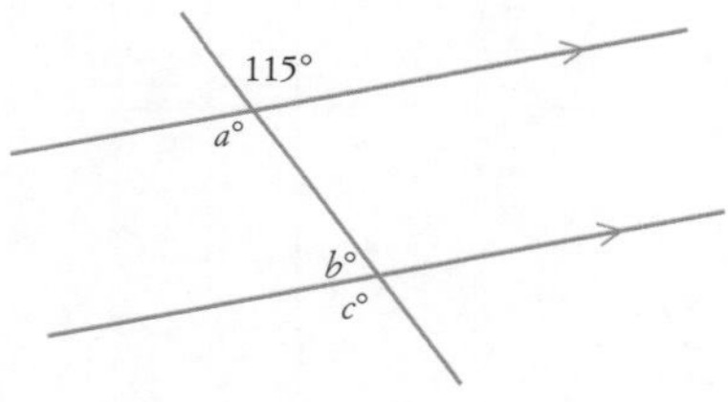

____

____

____

3 (3 marks) All 8 angles in this diagram are either x or y in size. 2 angles are marked. Complete the other 6 angles with an x or y.

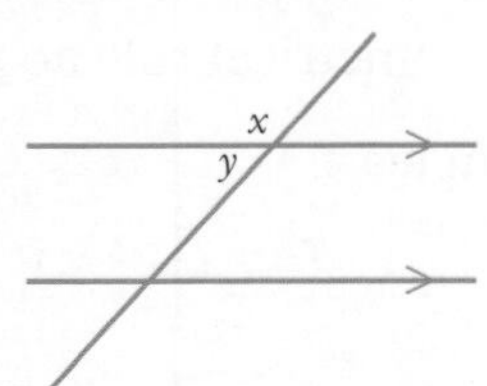

PART D: NUMERACY AND LITERACY

1 **a** Mark a pair of cointerior angles.

b How many different pairs of cointerior angles are there in total? ____________

c Describe what cointerior angles are.

____

____

____

d Complete this rule: Cointerior angles ______ parallel ____________ are ____________.

2 **a** Mark the angle that is corresponding to.

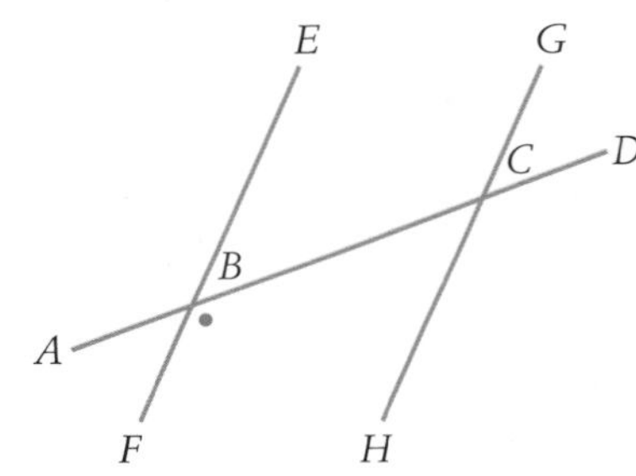

b Name this angle using 3 letters. ____________

3 Find the value of m and p, giving reasons.

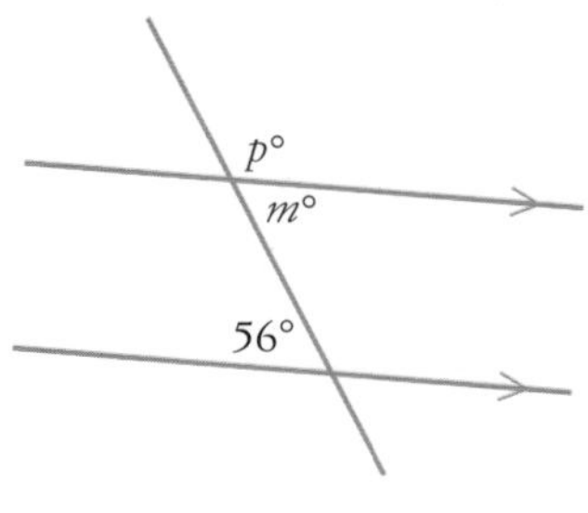

____

____

____

9780170454452

Name:

Due date:

Parent's signature:

Part A	/ 8 marks
Part B	/ 8 marks
Part C	/ 8 marks
Part D	/ 8 marks
Total	/ 32 marks

ANGLES REVISION 

HAVE YOU WORKED OUT WHY SOME QUESTIONS ARE BLUE AND SOME ARE GREEN? THERE'S EVEN A RED ONE HERE. YOU'LL NEED MORE TIME TO DO THAT ONE.

PART A: MENTAL MATHS

Calculators not allowed

1 Evaluate 6.2×2.

2 Complete: 3420 g = ________ kg.

3 What is a rectangle?

4 Evaluate $121 \div 11$.

5 Write 1:45 p.m. in 24-hour time.

6 Find the area of this triangle.

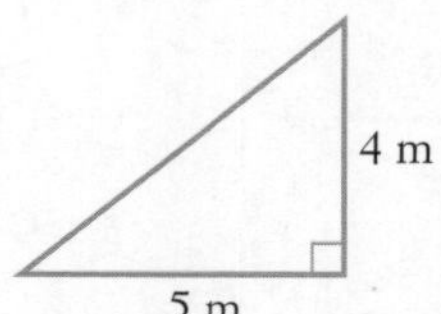

7 Find the average of 8, 7, 10, 9 and 1.

8 Evaluate $12 \div 3 + 5 \times 2$.

PART B: REVIEW

1 Find the value of each variable.

a

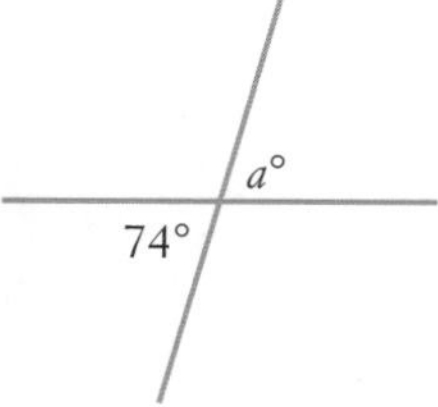

b

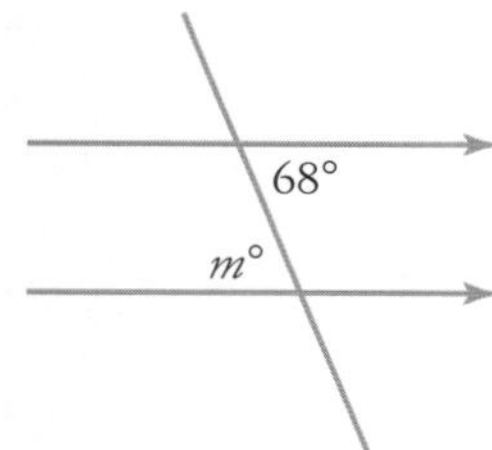

2 Draw a line through P perpendicular to the line below.

• P

C
S
F

9780170454452

3 (4 marks) Find the value of each variable, giving reasons.

a

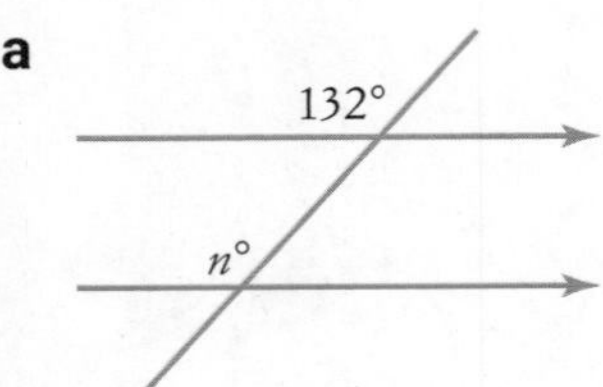

b

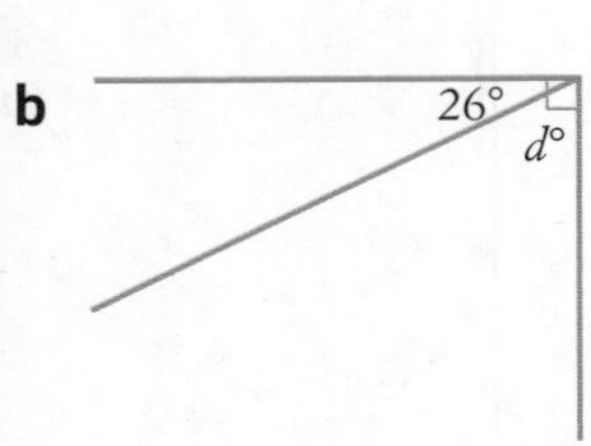

4 Construct an 88° angle.

PART C: PRACTICE

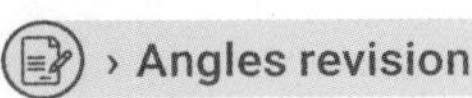

1 Complete: 2 angles that add up to 180° are called ______________________ angles.

2 **a** What is this shape called?

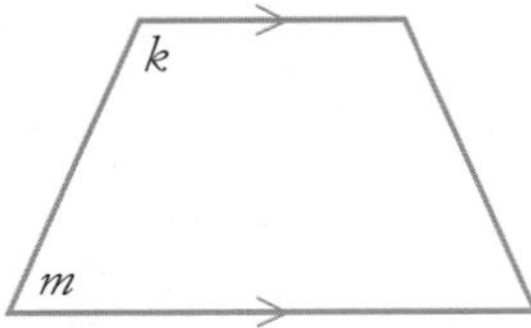

____

b Complete: Angles k and m are ____________ angles on parallel lines.

c If k is 144°, what is the size of m?

____

3 Find the value of a and b.

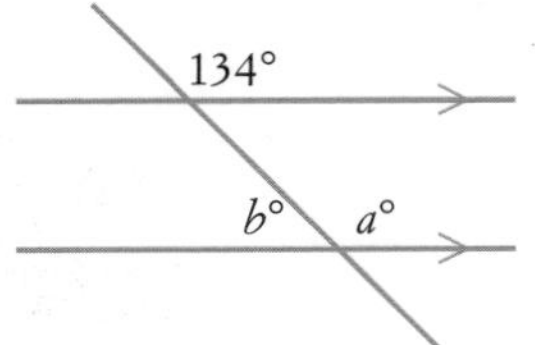

____

4 (2 marks) Are lines AB and CD parallel? Give a reason for your answer.

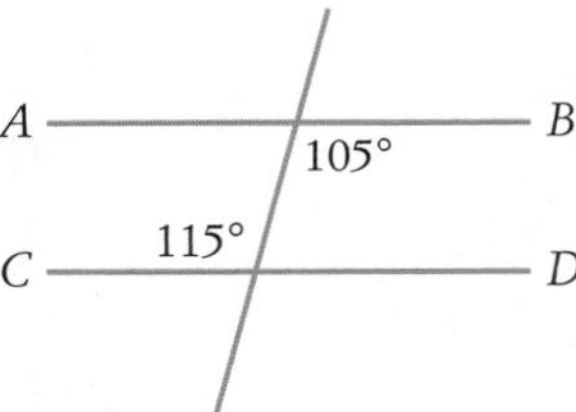

____

____

9780170454452

PART D: NUMERACY AND LITERACY

1 **a** What are complementary angles?

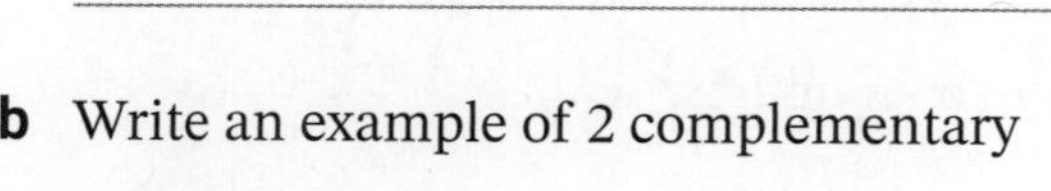

b Write an example of 2 complementary angle sizes.

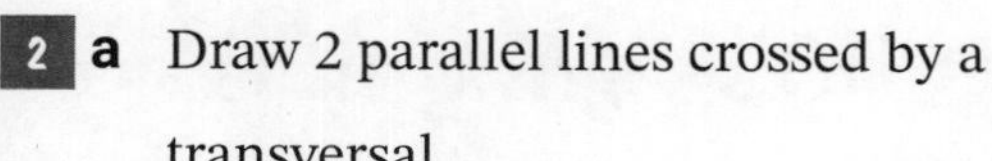

2 **a** Draw 2 parallel lines crossed by a transversal.

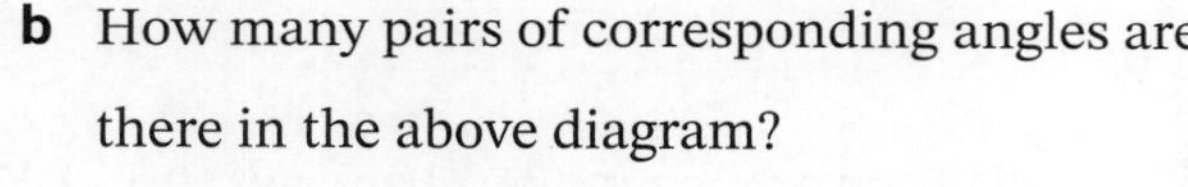

b How many pairs of corresponding angles are there in the above diagram?

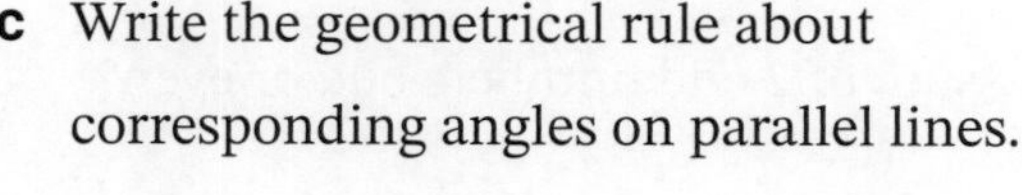

c Write the geometrical rule about corresponding angles on parallel lines.

3 (2 marks) Find the value of d, giving reasons.

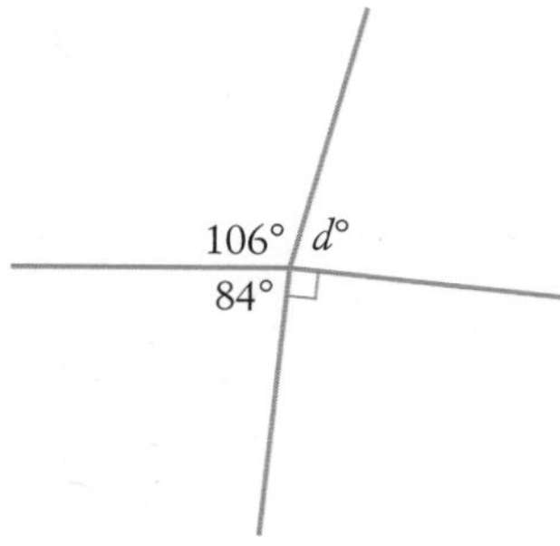

4 If 2 obtuse angles are combined, what type of angle is formed?

HW HOMEWORK

WS WORKSHEET

3 STARTUP ASSIGNMENT 3

PART A: BASIC SKILLS

/ 15 marks

1 How many metres are there in $\frac{1}{2}$ km? ______

2 Evaluate:

a $7 \times 2 \times 5$ ______

b 3.142×100 ______

c $\$30.00 - \$26.85$ ______

3 Simplify $\frac{12}{40}$. ______

4 Write 736 241 correct to the nearest thousand.

______

5 How many degrees in a revolution? ______

6 Find the average of 17, 11, 20 and 16.

______

7 What is the value of the 7 in 376 012?

______

8 Find the perimeter of this square.

______

9 Write 6:15 p.m. in 24-hour time. ______

10 How many possible outcomes are there if a letter is selected from the alphabet? ______

11 Write $\frac{3}{4}$ as a decimal. ______

12 How many days are there in October? ______

13 What type of triangle has 2 equal sides?

______

PART B: NUMBER

/ 25 marks

14 Write twelve thousand, seven hundred and forty-six in figures.

______

15 Write these numbers in ascending order.

7352, 10 840, 5216, 9004

______

16 Evaluate:

a $46 \div 2$ ______

b $5 \times 5 \times 5$ ______

c 3^2 ______

d 17×9 ______

17 List all of the prime numbers between 10 and 20.

______

18 Is the sum of 2 odd numbers odd or even?

______

19 Find the sum of the first 10 positive integers.

______

______

20 Insert grouping symbols (brackets) to make this equation true.

$8 + 10 - 3 \times 3 + 6 = 3$

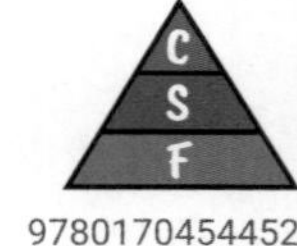

9780170454452

21 True or false?

a $4 \times 5 = 5 \times 4$ ____________

b $4 - 5 = 5 - 4$ ____________

c $12 > 23$ ____________

d $16 - 4 \times 2 + 1 < 10$ ____________

22 Look at the number 461 029.

a Write it in words.

____

____

____

b Which digit is in the hundreds place? ______

23 Write 2 numbers that have a product of:

a 45 ____________

b 100 ____________

24 Complete: If a number ends with the digit 0 or 5, then it must be divisible by ______.

25 If $\square^2 = 16$, what is the value of $\square$?

____

26 Write the number between 45 and 50 that is a multiple of 7. ____________

27 Evaluate:

a 9×8 ____________

b $176 \div 8$ ____________

c $18 + 12 \div 3 - 2$ ____________

d $75 \div (18 - 13) \times 10$ ____________

PART C: CHALLENGE

Bonus / 3 marks

Complete this magic square so that the numbers in every row, column and diagonal add up to 34. All of the numbers from 1 to 16 should appear.

1		14	
	6		
		2	16

WS WORKSHEET

C
S
F

3 POWERS AND ROOTS

INDEX IS ANOTHER WORD FOR POWER. POWERS ARE AN ABBREVIATION FOR REPEATED MULTIPLYING.

WORKSHEET

1 Use index notation to write each product.

a $5 \times 5 \times 5 \times 5$

b $2 \times 2 \times 2 \times 2 \times 2 \times 2 \times 2$

c $7 \times 7 \times 7$

d 16×16

e $3 \times 3 \times 3 \times 3 \times 3 \times 3 \times 3 \times 3$

f $2.5 \times 2.5 \times 2.5 \times 2.5 \times 2.5$

g $(-9) \times (-9) \times (-9)$

h $\frac{2}{3} \times \frac{2}{3} \times \frac{2}{3} \times \frac{2}{3} \times \frac{2}{3} \times \frac{2}{3}$

2 Write in expanded form.

a 3^4

b 5^2

c 7^5

d 11.2^4

e $(-8)^7$

f $(1\frac{1}{2})^3$

3 Evaluate each power.

a 13^2

b 9^3

c 8^2

d 5^3

e $(-2)^4$

f $(-10)^2$

g 2^7

h 10^6

i 1^{12}

j $(1.2)^2$

k 21^4

l $(-3)^7$

m 21^2

n $(-5)^5$

o $(2.3)^3$

p 3^{10}

4 Evaluate each root.

a $\sqrt{49}$

b $\sqrt{169}$

c $\sqrt{961}$

d $\sqrt{7.84}$

e $\sqrt{1}$

f $\sqrt{32\ 400}$

g $\sqrt[3]{729}$

h $\sqrt[3]{8}$

i $\sqrt[3]{-8}$

j $\sqrt[3]{4.913}$

k $\sqrt[3]{-4.913}$

l $\sqrt[3]{-0.216}$

5 Evaluate each root, correct to 2 decimal places.

a $\sqrt{40}$

b $\sqrt{75}$

c $\sqrt[3]{90}$

d $\sqrt[3]{-12}$

C S F

9780170454452

6 Is it possible to find:

a the square of a negative number?

b the cube of a negative number?

c the square root of a negative number?

d the cube root of a negative number?

e the square root of 0?

f the cube of 0?

7 Evaluate each power of 2.

a 2^2

b 2^3

c 2^4

d 2^5

e 2^6

f 2^7

g 2^8

h 2^9

8 Evaluate each power of 10.

a 10^2

b 10^6

c 10^9

d 10^7

e 10^3

f 10^4

g 10^{10}

h 10^5

9 What is the value of 10^{14}?

10 **a** If $18^2 = 324$, then what is $\sqrt{324}$?

b If $\sqrt{256} = 16$, then what is 16^2?

c If $7^3 = 343$, then what is $\sqrt[3]{343}$?

d If $\sqrt[3]{1331} = 11$, then what is 11^3?

11 Find the missing power.

a $2^? = 32$

b $3^? = 729$

c $7^? = 16\,807$

d $5^? = 625$

WS WORKSHEET

C S F

3 FACTOR TREES

Complete each factor tree and write each number as a product of its prime factors.

1
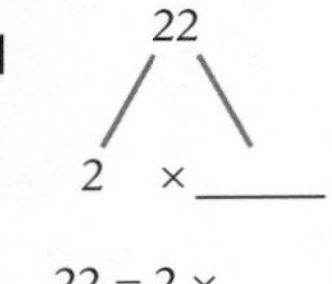

$22 = 2 \times$ ____

2

$38 =$ ____ $\times 2$

3
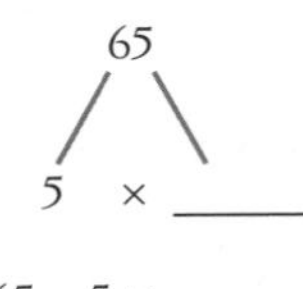

$65 = 5 \times$ ____

4
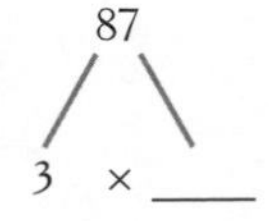

$87 = 3 \times$ ____

5

$146 = 2 \times$ ____

6
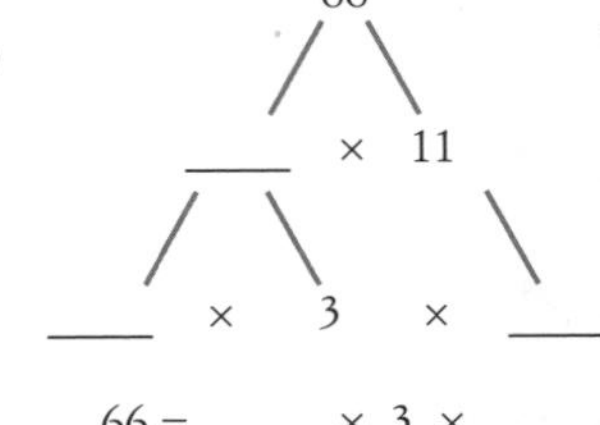

$66 =$ ____ $\times 3 \times$ ____

7
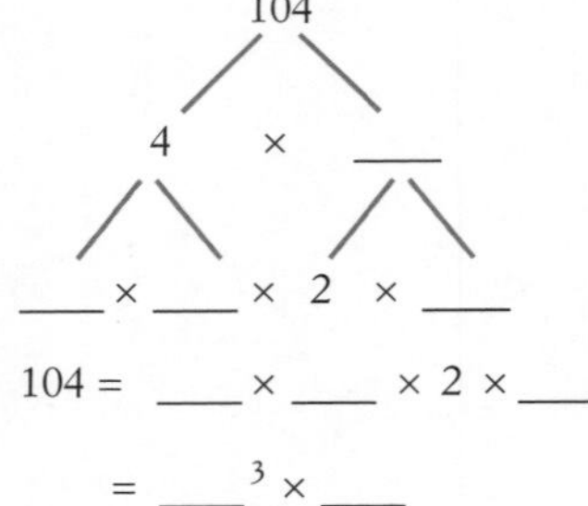

$104 =$ ____ $\times$ ____ $\times 2 \times$ ____

$=$ ____$^3 \times$ ____

8
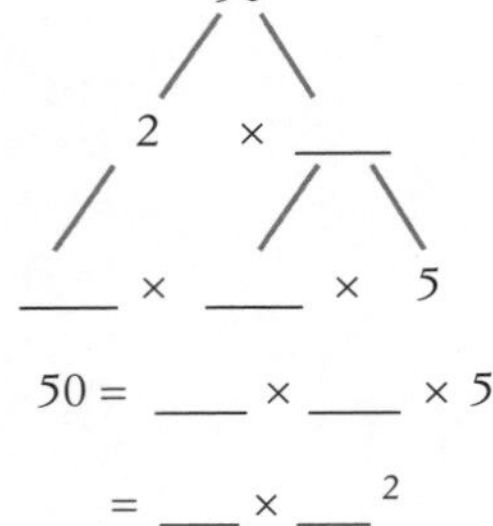

$50 =$ ____ $\times$ ____ $\times 5$

$=$ ____ $\times$ ____^{2}

9
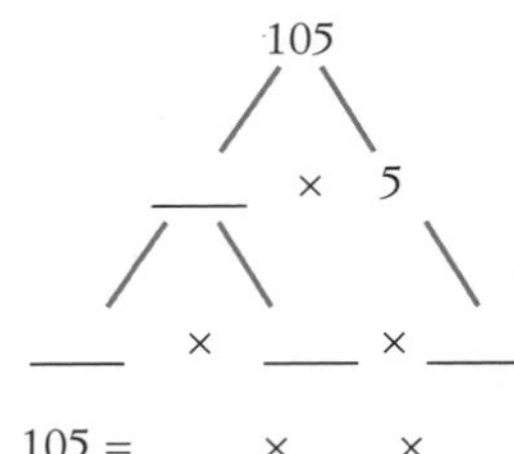

$105 =$ ____ $\times$ ____ $\times$ ____

10
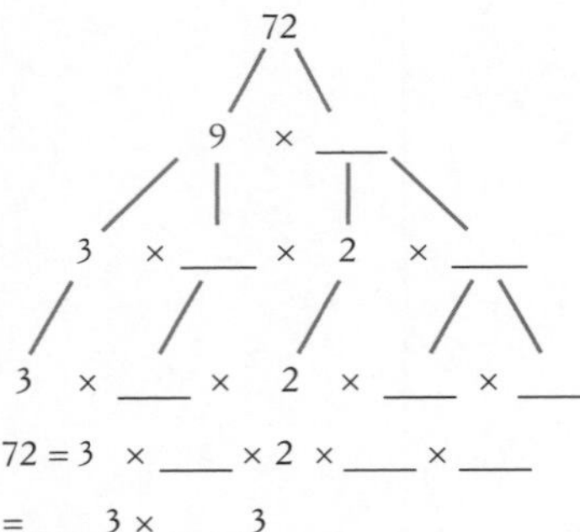

$72 = 3 \times$ ____ $\times 2 \times$ ____ $\times$ ____

$=$ ____$^3 \times$ ____^{3}

11
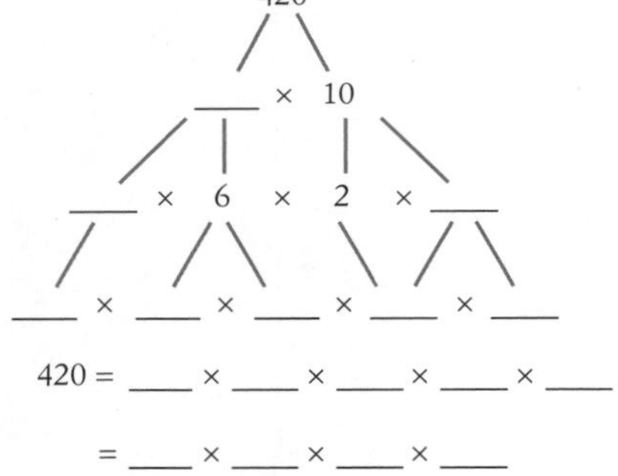

$420 =$ ____ $\times$ ____ $\times$ ____ $\times$ ____ $\times$ ____

$=$ ____ $\times$ ____ $\times$ ____ $\times$ ____

12
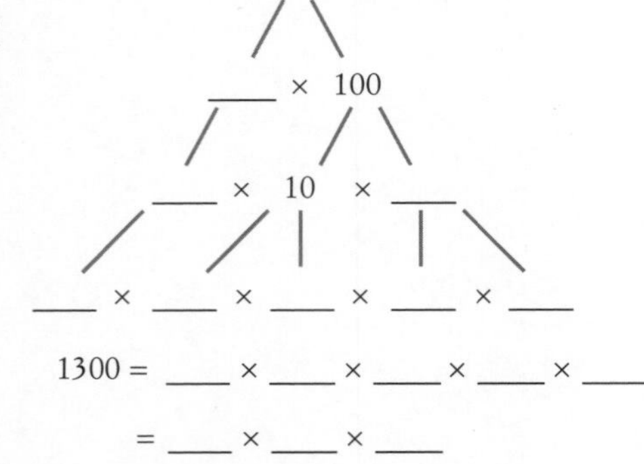

$1300 =$ ____ $\times$ ____ $\times$ ____ $\times$ ____ $\times$ ____

$=$ ____ $\times$ ____ $\times$ ____

13
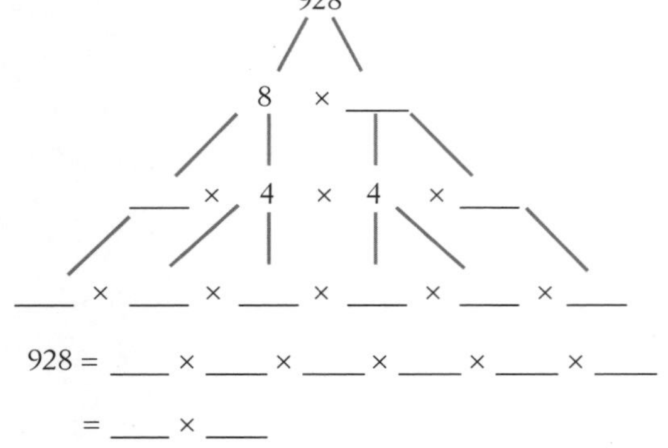

$928 =$ ____ $\times$ ____ $\times$ ____ $\times$ ____ $\times$ ____ $\times$ ____

$=$ ____ $\times$ ____

9780170454452

NUMBER FIND-A-WORD 

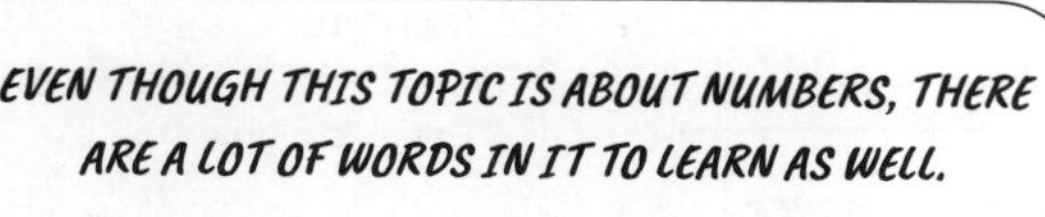


```
E C A N I S N F T B H V J W L E T A M I T S E S Q
N J T B Y O E H X I B R A O S S W Q U O T I E N T
F S E D I J O A R U N R B L R Z Q H S H X S M O X
M D R L F U M N Q X I D A F U D T U L K U J A M G
Z I L H S T R R P A Y H E C Q E L N A B B P K M D
N I L A W O R D Q S M F E X E Z S S T R P V M O Z
B U N L S D I V I S I B I L I T Y R T R E Q C C C
R D M I I W F O M E C A L P E I A U O F D D M I E
U C V E I O R N R C R F D W G C T X Q E M V B V K
V I Y V R E N U R N F T N I T S I Y B S W A A B E
D P Y C D A A M E E K R U I G M N U T H R L N N S
I R P R R Y L B T R Y E O A A U C E U A U Q O R Y
R O O T G A K E A E D N R T N S K D U A G I I T S
G D M D B Z X R E F U R E C C C I D T B T B T W T
C U P E F P U M R F M F I M A G N E H I X K A I E
N C G T A Y I H G I U X U R E W O P D N D S T D M
A T E N X C S Z M D K S B E H S N D O W S L O W I
Z N D S E S E H T N E R A P H G A T M E B C N Q R
E E T I S O P M O C E L P I T L U M L Z D O A M P
D F L O P E R A T I O N S C G R O T C A F G Z G O
```

Find these words in the puzzle above. They are across, up and down, and diagonal, and can be backwards as well as forwards.

ADDITION	APPROXIMATE	BILLION	BRACKETS
COMMON	COMPOSITE	CUBED	DIFFERENCE
DIVISIBILITY	DIVISOR	ESTIMATE	EVALUATE
EXPANDED	FACTOR	GREATER	INDEX
LESS	MILLION	MULTIPLE	NOTATION
NUMBER	OPERATIONS	ORDER	PLACE
POWER	PRIME	PRODUCT	QUOTIENT
ROOT	ROUND	SQUARED	SUM
THOUSAND	VALUE		

3 WHOLE NUMBERS

THIS ASSIGNMENT COVERS MENTAL MATHS, MULTIPLYING AND DIVIDING NUMBERS, AND DIVISIBILITY TESTS.

Name:

Due date:

Parent's signature:

Part A	/ 8 marks
Part B	/ 8 marks
Part C	/ 8 marks
Part D	/ 8 marks
Total	/ 32 marks

HW HOMEWORK

PART A: MENTAL MATHS

Calculators are not allowed

1 Simplify $\frac{18}{30}$. ______

2 Evaluate \$100 − \$87.65. ______

3 List the factors of 25.

______

4 A movie starts at 2:45 p.m. and runs for 135 minutes. What time does it end?

______

5 What is the next number in this pattern?

1, 3, 9, 27, ______

6 Complete: 9.35 km = ______ m.

7 Find the perimeter of this figure.

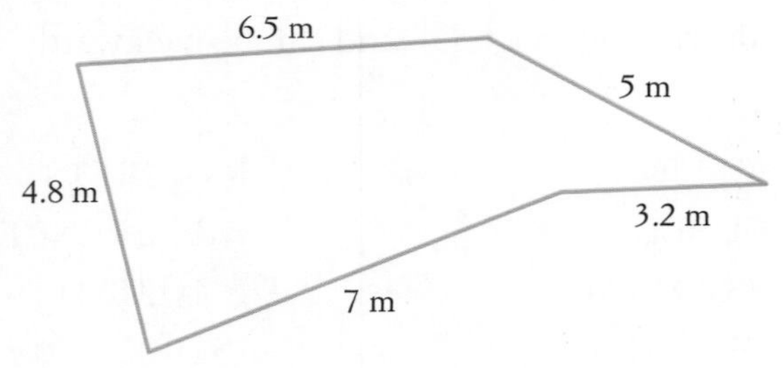

______

8 Evaluate 160 ÷ 4 + 8 × 5.

______

PART B: REVIEW

1 Evaluate each expression by mental calculation (without calculators).

a 6 × 8 ______

b 2 × 98 × 5 ______

c 7400 ÷ 10 ______

2 Is 456 divisible by 3?

______

3 Evaluate each expression.

a 74 × 8

______

______

b 1260 × 20

______

______

c 91 ÷ 7

______

______

d 540 ÷ 20

______

______

C S F

9780170454452

PART C: PRACTICE

› Rounding and estimating
› Multiplying and dividing numbers
› Divisibility tests

1 Round 209 415 to the nearest:

a ten ______

b thousand ______

2 Estimate the answer to:

145 + 89 + 23 + 160 + 37.

______

3 Test whether 1010 is divisible by:

a 2

______

b 4

______

c 6

______

4 Evaluate 198 ÷ 9.

______

5 If a number is divisible by 3 and 5, what other number must it be divisible by?

______

PART D: NUMERACY AND LITERACY

1 Jai can wash 7 cars in one hour. How many hours will it take him to wash 100 cars? Answer to the nearest hour.

______

______

2 Evaluate each expression by mental calculation and explain how you found the answer.

a 16×5

______

______

______

______

b 28×100

______

______

3 **a** In the number 16 379, which digit is in the hundreds place? ______

b If rounding 16 379 to the nearest hundred, explain why we round up.

______

______

______

______

c Round 16 379 to the nearest hundred.

______

4 Is 468 is divisible by each number below? Give a reason involving a divisibility test.

a 5

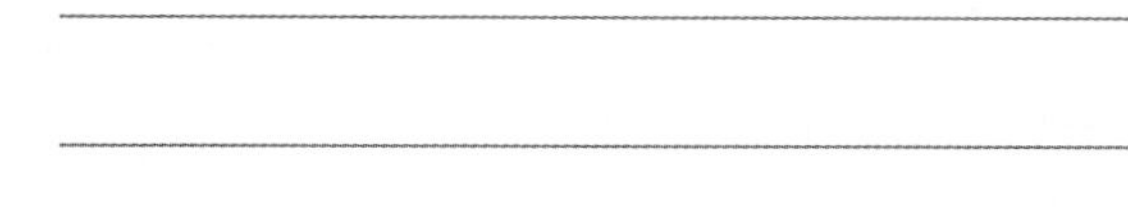

b 9

3 POWERS AND SQUARE ROOTS

Name:

Due date:

Parent's signature:

Part A	/ 8 marks
Part B	/ 8 marks
Part C	/ 8 marks
Part D	/ 8 marks
Total	/ 32 marks

SOME CLUES FOR THIS ASSIGNMENT: THERE ARE 6 ZEROES IN A MILLION; TO DIVIDE BY 5, DIVIDE BY 10 AND DOUBLE.

PART A: MENTAL MATHS

Calculators not allowed

1 List the first 5 multiples of 8.

2 What type of angle is less than 90°?

3 Evaluate $2 \times 2 \times 2 \times 2 \times 2$.

4 Find the area of this rectangle.

5.2 m

3 m

5 Write 04:15 in 12-hour (a.m./p.m.) time.

6 Evaluate $654 \div 6$.

7 Write 70% as a simple fraction.

8 Complete: 4120 mm = ____________ cm.

PART B: REVIEW

1 Round 3 751 940 to the nearest ten thousand.

2 (2 marks) Test whether 1824 is divisible by 6, giving reasons for your answer.

3 Evaluate each expression by mental calculation.

a 11×11

b 45×40

c $310 \div 5$

d $272 \div 4$

4 Estimate the value of 72×12, giving reasons for your answer.

C S F

HW HOMEWORK

9780170454452

PART C: PRACTICE

› Powers and index notation
› Square and square root

1 Write each expression using index notation.

a $3 \times 3 \times 3 \times 3 \times 3$ ________________

b $5 \times 5 \times 7 \times 7 \times 7 \times 7$ ________________

2 Evaluate each expression.

a 6^2 ________________

b 8^3 ________________

c $\sqrt{81}$ ________________

d $\sqrt{10^2}$ ________________

3 **a** Write 5^4 in expanded form.

____

b Evaluate 5^4 ________________

PART D: NUMERACY AND LITERACY

1 (2 marks) Complete: In 12^5, 12 is called the ________________ and 5 is called the ________________.

2 Explain what 12^5 means.

____

____

3 **a** Write **ten million** in figures.

____

b Complete the power of ten:

ten million = $10^{\square}$

4 **a** Explain what the **square root** of 441 means.

____

____

b Evaluate $\sqrt{441}$.

____

5 Find a square number between 60 and 70.

____

HW HOMEWORK

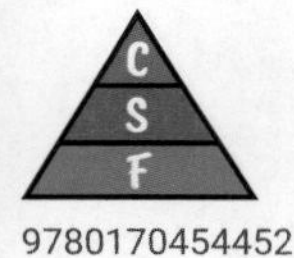

3 PRIME AND COMPOSITE NUMBERS

Name:

Due date:

Parent's signature:

Part A	/ 8 marks
Part B	/ 8 marks
Part C	/ 8 marks
Part D	/ 8 marks
Total	/ 32 marks

HW HOMEWORK

PART A: MENTAL MATHS

Calculators are not allowed

1 What is the name of a shape that has 6 sides?

2 Complete: The size of an obtuse angle is between ______ ° and ______ °.

3 Write 5:34 p.m. in 24-hour time. ______

4 Evaluate $3.2 - 0.06$.

5 How many days in December? ______

6 Find the product of 7 and 14.

7 Complete: 6.8 L = ______ mL.

8 Find the perimeter of this rhombus.

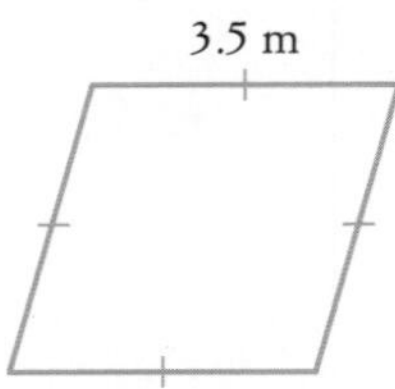

PART B: REVIEW

1 Round 1084 to the nearest hundred. ______

2 (2 marks) List the factors of 30.

3 Find 2 integers that have a product of 48.

4 Evaluate 1^5. ______

5 Is 1645 divisible by 6? ______

6 List the factors of 11.

7 Find 2 numbers that have a product of 21.

C S F

9780170454452

PART C: PRACTICE

› Prime and composite numbers
› Prime factors
› Highest common factor and lowest common multiple

1 List all of the prime numbers between 40 and 50.

____

2 (3 marks) Draw a factor tree for 84 and write 84 as a product of its prime factors.

3 (2 marks) Draw a factor tree for 60.

4 For 84 and 60, use the factor trees above to find:

a the highest common factor

____

b the lowest common multiple

____

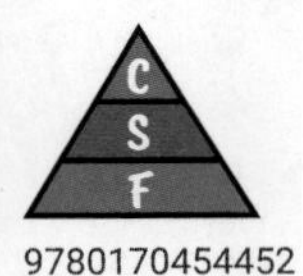

PART D: NUMERACY AND LITERACY

1 Explain how you used the factor trees to find the highest common factor of 84 and 60 in question **4a** of Part C.

____

____

____

2 What is a composite number?

____

____

3 Which whole number is neither prime nor composite? ______________________

4 (2 marks) Complete: A prime number has exactly 2 factors: ____________ and ____________.

5 What is the name given to the largest number that is a factor of 2 or more given numbers?

____

6 What is the smallest number that is divisible by both 6 and 8?

____

7 Find the multiple of 18 that is between 70 and 90.

____

HW HOMEWORK

3 WHOLE NUMBERS REVIEW

WE'RE NEAR THE END OF THE WHOLE NUMBERS TOPIC NOW. SOLVE ALL THESE PROBLEMS, THEN CONSIDER YOURSELF A MATHS MASTER!

Name:

Due date:

Parent's signature:

Part A	/ 8 marks
Part B	/ 8 marks
Part C	/ 8 marks
Part D	/ 8 marks
Total	/ 32 marks

HW HOMEWORK

PART A: MENTAL MATHS

Calculators are not allowed

1 Evaluate $9 \times 6 - 3 \times 2$.

2 How many seconds in one hour?

3 Find the average of 8, 18, 15 and 11.

4 If a rectangle has width of 8 cm and area of 56 cm^2, what is its length?

5 Write 643 589 in words.

6 What type of triangle has no equal sides?

7 Evaluate $180 \div 4$.

8 Evaluate $180 \div 20$.

PART B: REVIEW

1 Write $3 \times 3 \times 3 \times 3 \times 3 \times 4 \times 4 \times 4$ using index notation.

2 Evaluate 7^4.

3 Find the lowest common multiple of 16 and 20.

4 (2 marks) List all the prime numbers between 20 and 40.

5 Evaluate $\sqrt{169}$.

6 Test whether 294 is divisible by 6, showing working.

7 Find the highest common factor of 28 and 36.

C S F

9780170454452

PART C: PRACTICE

› Whole numbers review

1 Evaluate 18^3.

2 Evaluate 256×22.

3 How can you tell that 4084 is not divisible by 5?

4 (3 marks) Draw a factor tree for 72 and write 72 as a product of its prime factors.

5 If 96 as a product of its prime factors is $2 \times 2 \times 2 \times 2 \times 2 \times 3$, find the highest common factor of 72 and 96

6 Write a number that can be rounded to 16 000. ______

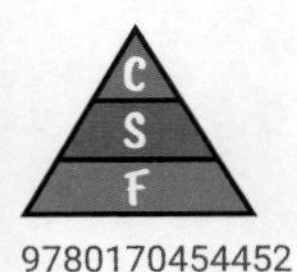

PART D: NUMERACY AND LITERACY

1 Is 9 prime or composite?

2 **a** Write '8^3' in words.

b Write 8^3 in expanded form.

c In 8^3, what is the base? ______

3 **a** Estimate the value of 28×9. Show working.

b Evaluate 28×9 by mental calculation. Show working.

4 (2 marks) Complete: The ______ ______ of 49 is ______ because $7^2 =$ ______.

HW HOMEWORK

3 CROSS NUMBER PUZZLE

1		■	2			3	■	4	5
	■	6		■	■	7	8	■	
	■	9		10	■	11		■	
■	12	■	■		■		■	13	
14			15		■	16	17		
■		■		■	18	■	19		■
20		21	■	■	22	23		■	24
	■	25	26	■	27		■	■	
	■	28		■	■	29		30	
31		■	32				■	33	

Across

1 $2 \times 2 \times 3 \times 3$

2 $10 \times 10 \times 10 + 1$

4 Seven times six divided by two

6 $768 \div 12$

7 $3 \times 3 \times 3$

9 $213 + 2 + 127$

11 $6805 - 6773$

13 $100 - 16$

14 Eleven thousand and eleven plus one hundred and forty-four

16 $10\,000 - 4765$

19 $891 \div 11$

20 Four hundred and forty-four

22 $200 - 2$

25 $110 \div 10$

27 7^2

28 Number of eggs in a dozen

29 Largest number using 6, 8, 9, 7

31 3×6

32 $1300 - 81$

33 $3\,000\,000 \div 100\,000$

Down

1 Number of days in a leap year

2 12×12

3 Smallest number using 3, 2, 4, 1, 5

5 The first 5 positive integers

6 $3 \times 3 \times 7$

8 9×8

10 $600 - 365$

12 $4496 \div 4$

13 $8 \times 100 + 3 \times 10 + 1$

15 $2 \times 3 \times 3 \times 3$

17 144×2

18 $2 + 2 + 8 + 9 + 32 + 25 + 36$

20 $12\,273 \div 3$

21 137×3

23 One less than ten thousand

24 $8800 \div 5$

26 $3025 \div 25$

30 $7 \times 9 + 10$

STARTUP ASSIGNMENT 4 (4)

SOME STUDENTS HAVE TROUBLE WITH FRACTIONS AND PERCENTAGES. HERE ARE SOME SKILLS TO PRACTISE TO HELP YOU LEARN THIS TOPIC BETTER.

WS WORKSHEET

PPART A: BASIC SKILLS

/ 15 marks

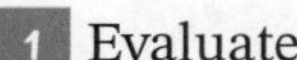

1 Evaluate:

a 35×100 ______

b 23×11 ______

c 0.5×40 ______

d $384 \div 4$ ______

2 A rectangle has length 8 cm and width 3 cm.
What is its area? ______

3 Write $\frac{1}{4}$ as:

a a decimal ______

b a percentage ______

4 Find y.

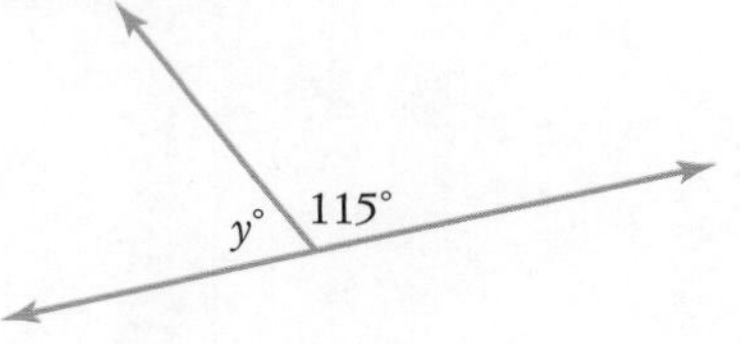

______

5 Write 240 176 correct to the nearest hundred.

______

6 Evaluate:

a $-3 + 4 + (-5)$ ______

b $40 - 12 \div 2$ ______

c 16^2 ______

7 Use index notation to write $9 \times 9 \times 9 \times 9$.

______

8 Complete: 400 mL = ______ L

9 Name this shape.

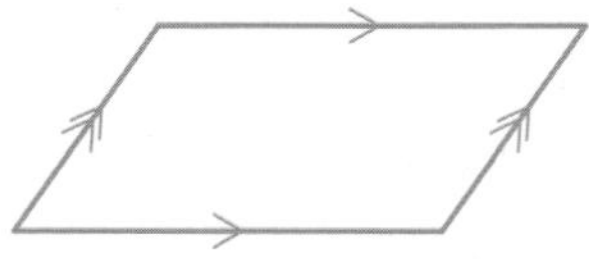

______

PART B: NUMBER

/ 25 marks

10 Find:

a $27 \div 3$ ______

b 4×8 ______

c 15×2 ______

d $28 \div 4$ ______

11 List the first 5 multiples of 8.

______

12 What fraction of this diagram is shaded?

______

13 Convert each percentage to a simplified fraction.

a 67% ______

b 20% ______

14 List the factors of:

a 16

______

b 45

______

15 Is 22 divisible by 3?

____

16 Complete

a ________ $\times 3 = 12$

b ________ $\times 7 = 35$

17 What is the highest common factor of 18 and 30?

____

18 At the party, there were 16 boys and 11 girls.

What fraction of the total were girls?

______________________.

19 Complete:

a 4 hours = ________ minutes

b 2 kg = ________ g

c 240 cm = ________ mm

d 3 days = ________ hours

20 Find:

a $35 \div 7$ ________

b $\frac{1}{3} \times 21$ ________

c $\frac{1}{2} \times 32$ ________

d $\frac{1}{5} \times 30$ ________

21 Is 12 a multiple of 3? ________

22 What is the lowest common multiple of 20 and 8?

____

PART C: CHALLENGE

Bonus / 3 marks

Mandeep sold all of her eggs.

She first sold half her eggs plus half an egg to her neighbour, without needing to break an egg.

Then she sold half her remaining eggs plus half an egg to her uncle, again without breaking an egg.

Finally, she sold half her remaining eggs plus half an egg to her friend, without breaking an egg.

How many eggs did Mandeep begin with?

C S F

9780170454452

PERCENTAGES CROSS NUMBER 

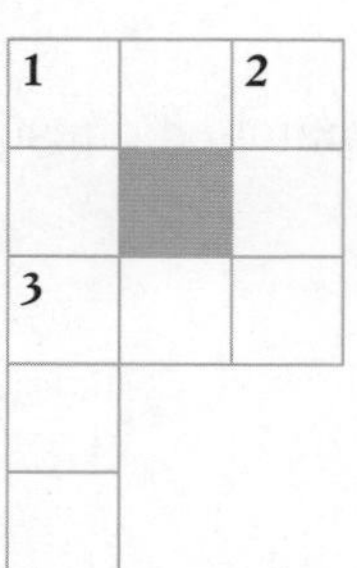

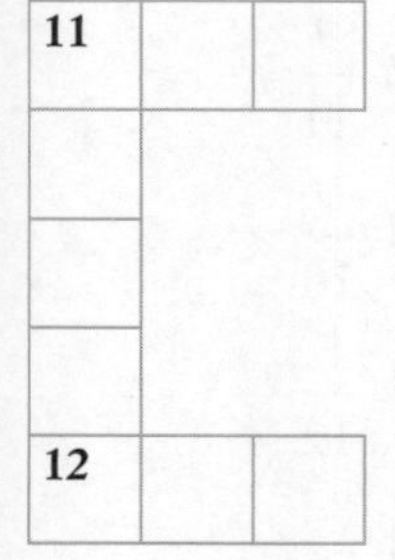

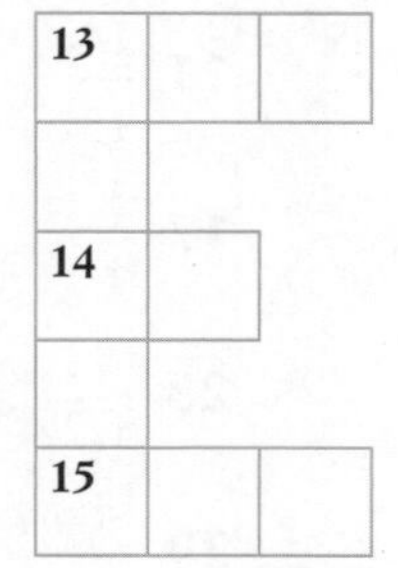

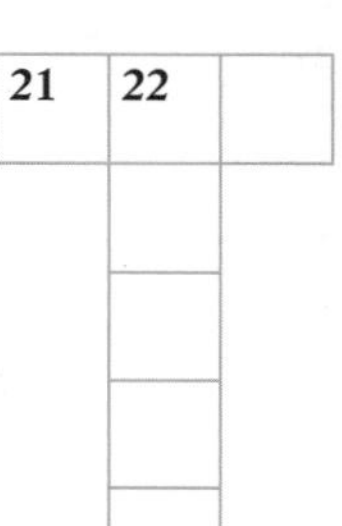

PUZZLE SHEET

Across

1 4% of 3800

3 5% of 13 000

4 3% of 12 700

5 42% of 150

6 16% of 5312.5

7 120% of 185

9 7% of 500

11 20% of 1400

12 9% of 8900

13 5% of 9760

14 22% of 400

15 50% of 200

17 30% of 270

19 30% of 170

20 130% of 40

21 24% of 2425

Down

1 10% of 126 300

2 8% of 2625

4 25% of 142 472

7 80% of 25 457.5

8 40% of 70

10 $12\frac{1}{2}$% of 600

11 25% of 107 352

13 19% of 214 900

16 40% of 72 385

18 75% of 14 816

19 1% of 5500

22 60% of 135 205

4 EQUIVALENT FRACTIONS

FOR THIS PUZZLE, YOU NEED TO PAIR UP EQUIVALENT FRACTIONS FROM LISTS A AND B.

PUZZLE SHEET

Match each fraction in List A to its equivalent fraction in List B. Use the matched question numbers and letters to decode the question next page, then answer the question.

List A

1 $\frac{3}{5}$	**2** $\frac{2}{3}$	**3** $\frac{4}{10}$	**4** $\frac{8}{16}$	**5** $\frac{1}{5}$	**6** $\frac{7}{10}$
7 $\frac{10}{12}$	**8** $\frac{3}{4}$	**9** $\frac{5}{7}$	**10** $\frac{1}{6}$	**11** $\frac{14}{15}$	**12** $\frac{9}{11}$
13 $\frac{1}{8}$	**14** $\frac{16}{20}$	**15** $\frac{2}{6}$	**16** $\frac{5}{8}$	**17** $\frac{3}{7}$	**18** $\frac{9}{10}$
19 $\frac{1}{11}$	**20** $\frac{13}{20}$	**21** $\frac{5}{30}$	**22** $\frac{1}{4}$	**23** $\frac{3}{16}$	**24** $\frac{5}{12}$
25 $\frac{21}{27}$	**26** $\frac{7}{8}$	**27** $\frac{15}{50}$	**28** $\frac{1}{7}$	**29** $\frac{4}{15}$	**30** $\frac{8}{9}$

List B

M $\frac{3}{10}$	**B** $\frac{15}{35}$	**A** $\frac{27}{33}$	**Z** $\frac{4}{24}$	**O** $\frac{35}{40}$	**T** $\frac{4}{6}$
S $\frac{18}{24}$	**R** $\frac{3}{11}$	**C** $\frac{28}{30}$	**X** $\frac{4}{8}$	**Y** $\frac{1}{6}$	**E** $\frac{10}{50}$
T $\frac{8}{88}$	**L** $\frac{5}{6}$	**I** $\frac{10}{24}$	**F** $\frac{40}{45}$	**U** $\frac{6}{42}$	**H** $\frac{9}{15}$
J $\frac{3}{24}$	**V** $\frac{1}{3}$	**G** $\frac{21}{30}$	**N** $\frac{2}{5}$	**E** $\frac{10}{16}$	**D** $\frac{3}{12}$
P $\frac{6}{32}$	**R** $\frac{8}{30}$	**K** $\frac{39}{60}$	**Q** $\frac{4}{5}$	**W** $\frac{10}{14}$	**A** $\frac{36}{40}$

9780170454452

9	1	12	19

24	8

2	1	5

3	18	27	16

26	30

2	1	16

7	24	3	5

19	1	12	2

8	5	23	18	29	12	19	5	8

2	1	5

3	28	27	16	29	12	2	26	29

30	25	26	27

19	1	5

22	16	3	26	27	24	3	18	19	26	29

26	30

18

30	29	12	11	19	24	26	3

?

4 FRACTIONS 1

Name:

Due date:

Parent's signature:

Part A	/ 8 marks
Part B	/ 8 marks
Part C	/ 8 marks
Part D	/ 8 marks
Total	/ 32 marks

ART A: MENTAL MATHS

Calculators are not allowed.

1 What type of angle is marked? ______

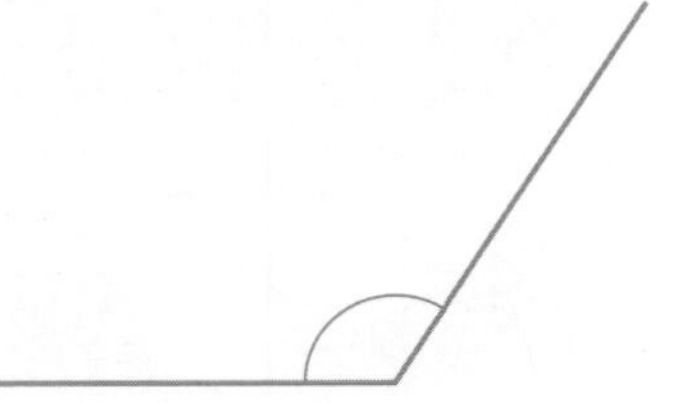

2 Evaluate 84 ÷ 3.

______

3 Write in ascending order: 7, −5, 0, −2, 1.

______

4 Complete: 9.41 t = ______ kg.

5 Evaluate 158 + 9 × 6.

______

6 Draw a trapezium.

7 List all the square numbers below 20.

______

8 Find the average of 18 and 26.

______

PART B: REVIEW

1 Shade $\frac{1}{2}$ of each shape.

a

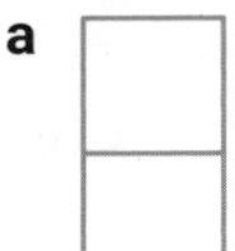

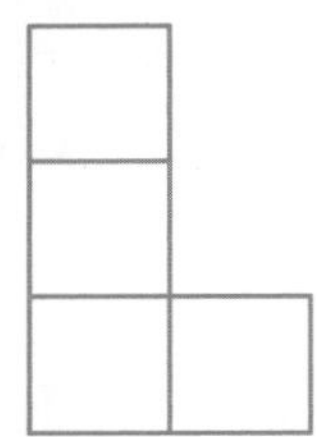

b

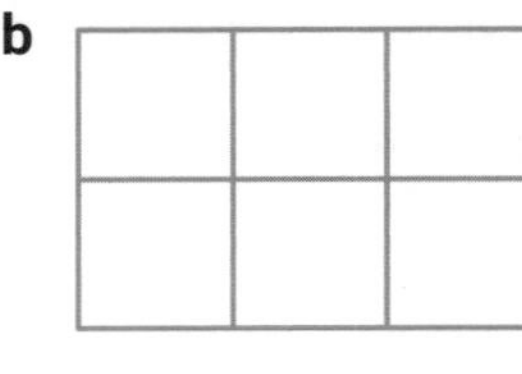

2 Find the lowest common multiple of:

a 2 and 3

______

b 8 and 10

______

3 Shade $\frac{3}{4}$ of each shape.

a

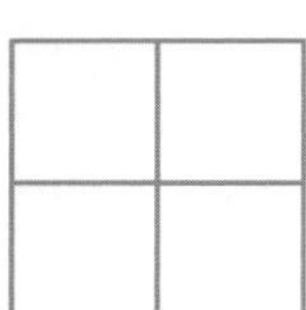

b

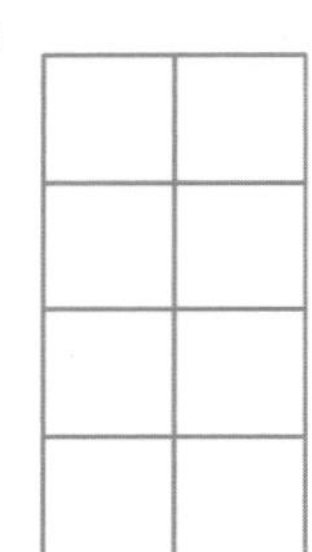

4 Find the highest common factor of:

a 24 and 18

______

b 12 and 20 ______

C S F

9780170454452

PART C: PRACTICE

› Simplifying fractions
› Ordering fractions

1 What is the number at the bottom of a fraction called (for example, the 10 in $\frac{3}{10}$)?

2 Which fraction is larger: $\frac{3}{5}$ or $\frac{4}{7}$? Show working.

3 Convert $4\frac{2}{3}$ to an improper fraction.

4 Complete:

a $\frac{3}{8} = \frac{}{24}$

b $\frac{5}{6} = \frac{35}{}$

5 Simplify each fraction and express as a mixed numeral if appropriate.

a $\frac{14}{24}$ ______

b $\frac{48}{20}$ ______

6 List these fractions in ascending order: $\frac{6}{10}, \frac{2}{5}, \frac{3}{4}$

C
S
F

PART D: NUMERACY AND LITERACY

1 **a** What is a proper fraction?

b Write an example of a proper fraction.

2 Write ‘$<$’ in words.

3 Complete:

a ‘Descending order’ means from ______ to ______

b To simplify a fraction, keep ______ its numerator and denominator by the same ______.

4 What type of fraction is $3\frac{2}{5}$?

5 (2 marks) Explain how to convert $\frac{23}{9}$ to a mixed numeral.

HW HOMEWORK

4 FRACTIONS 2

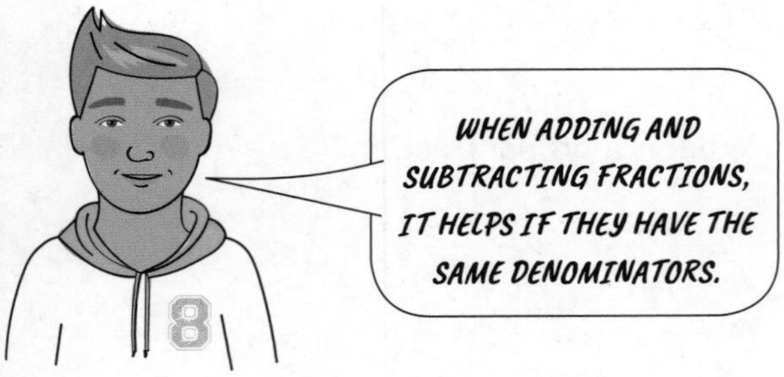

Name:

Due date:

Parent's signature:

Part A	/ 8 marks
Part B	/ 8 marks
Part C	/ 8 marks
Part D	/ 8 marks
Total	/ 32 marks

PART A: MENTAL MATHS

Calculators are not allowed.

1 Simplify 18×2.

____

2 How many sides has a pentagon? ____________

3 A movie started at 7:30 p.m. and ran for 1 hour and 54 minutes. At what time did it end?

____

4 Find the perimeter of this rectangle.

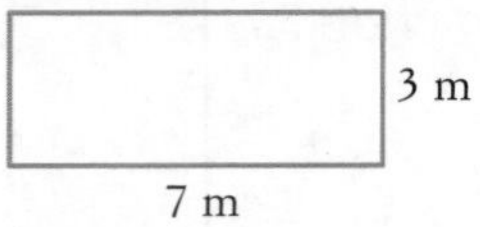

____

5 Find the average of 5, 8, 14, 6 and 12.

____

6 Complete: 9.6 kg = ____________ g.

7 Draw a right angle.

8 Evaluate 6×9. ____________

PART B: REVIEW

1 Simplify $\frac{15}{21}$.

____

2 Shade $\frac{2}{3}$ of this shape.

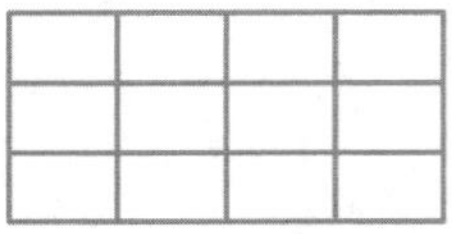

3 There were 18 boys and 22 girls at the student conference. What fraction were boys? Answer in simplest form.

____

4 Convert $\frac{17}{7}$ to a mixed numeral.

____

5 What fraction of one day is 8 hours?

____

6 Which fraction is larger: $\frac{7}{10}$ or $\frac{3}{4}$?

____

7 Write a fraction that simplifies to $\frac{1}{4}$.

____

8 List the first 4 multiples of 6.

____

C S F

9780170454452

PART C: PRACTICE

› Adding and subtracting fractions
› Fraction of a quantity

1 Evaluate and express as a mixed numeral if appropriate:

a $\frac{4}{9} - \frac{1}{9}$

____

b $\frac{2}{3} + \frac{2}{3}$

____

c $\frac{1}{5} + \frac{1}{4}$

____

2 Find $\frac{3}{5}$ of \$80.

____

3 Evaluate:

a $\frac{5}{6} - \frac{1}{2}$

____

b $\frac{7}{10} - \frac{1}{6}$

____

c $3\frac{1}{8} + 2\frac{3}{4}$

____

4 Find $\frac{2}{3}$ of 1 year (in months).

____

PART D: NUMERACY AND LITERACY

1 A prize of \$30 is divided between 3 people. Rina gets 50%, George gets $\frac{1}{6}$ and Jana gets what's left.

a How much does Rina get? ______________

b How much does George get? ______________

c What fraction does Jana get? ______________

2 A cake recipe requires $2\frac{1}{4}$ cups of flour and $\frac{1}{3}$ cup of sugar.

a How many cups are there altogther?

____

b How many cups of sugar are required to make 6 cakes?

____

3 What is an improper fraction?

____

4 What is the top number in a fraction called (for example, the 3 in $\frac{3}{10}$)?

____

5 Complete: To add or subtract fractions, we first convert them so that they have the same ______________

HW HOMEWORK

C S F

4 FRACTIONS 3

HAVE YOU MASTERED YOUR FRACTIONS SKILLS YET? THEY'RE NOT EASY, BUT THERE ARE NO SHORTCUTS. YOU JUST HAVE TO KEEP PRACTISING. GAME ON!

Name:

Due date:

Parent's signature:

Part A	/ 8 marks
Part B	/ 8 marks
Part C	/ 8 marks
Part D	/ 8 marks
Total	/ 32 marks

HW HOMEWORK

PART A: MENTAL MATHS

Calculators are not allowed.

1 Evaluate $3 - (-3)$.

2 What 4-sided shape has opposite sides equal and parallel?

3 List the factors of 20.

4 Convert 1000 cents to dollars.

5 Find the perimeter of this shape.

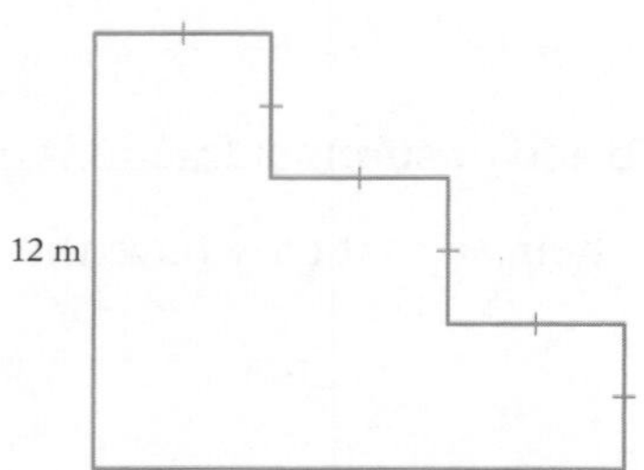

6 Evaluate $2 \times 4 \times 3 \times 5$.

7 Write 3:25 a.m. in 24-hour time.

8 Evaluate $1638 \div 6$.

PART B: REVIEW

1 Simplify $\frac{88}{100}$.

2 Is $\frac{7}{8} > \frac{5}{6}$? Show working.

3 Evaluate each expression.

a $\frac{3}{4} + 6\frac{2}{5}$

b $7 - 2\frac{3}{8}$

c $8\frac{4}{9} + 5\frac{1}{6}$

4 Complete: ______ $- \frac{2}{3} = \frac{5}{24}$

5 Find $\frac{2}{3}$ of 4 hours, in hours and minutes.

6 At a picnic, 30 students were given $\frac{1}{4}$ L of apple juice each. How many litres were given out?

C S F

9780170454452

PART C: PRACTICE

› Multiplying and dividing fractions
› Fraction problems

1 Convert $8\frac{3}{4}$ to an improper fraction.

2 Multiply any fraction by its reciprocal. What do you notice about the answer? Show working.

3 Evaluate:

a $\frac{7}{12} \times \frac{3}{14}$

b $1\frac{3}{4} \times 2\frac{1}{7}$

c $\frac{2}{3} \div \frac{4}{5}$

d $3\frac{1}{2} \div 2\frac{1}{3}$

e $\left(\frac{1}{9}\right)^2$

4 Complete: $\frac{5}{6} \times$ ______________ $= \frac{7}{12}$

C
S
F

PART D: NUMERACY AND LITERACY

1 Minjun takes 10 hours to paint a house. He painted for $3\frac{3}{5}$ hours on Saturday and $1\frac{1}{2}$ hours on Sunday. How many more hours does he need?

2 Complete: To divide by a fraction, multiply by its ______________

3 Jane divides $\frac{4}{5}$ of a pizza evenly between her 3 sisters. What fraction of the pizza does each sister receive?

4 (2 marks) Explain how to evaluate $\frac{9}{10} \times \frac{2}{3}$.

5 Ibrahim had 80 boxes of chocolate to sell. His family bought $\frac{1}{10}$ of them while his 4 friends bought $\frac{1}{16}$ each.

a How many boxes did each friend buy?

b How many boxes are left?

c What fraction of boxes are left?

HW HOMEWORK

4 PERCENTAGES

PERCENTAGES ARE SPECIAL FRACTIONS THAT HAVE A DENOMINATOR OF 100. FOR EXAMPLE, 50% MEANS $\frac{50}{100}$ OR $\frac{1}{2}$.

Name:

Due date:

Parent's signature:

Part A	/ 8 marks
Part B	/ 8 marks
Part C	/ 8 marks
Part D	/ 8 marks
Total	/ 32 marks

HOMEWORK

PART A: MENTAL MATHS

Calculators are not allowed.

1 Evaluate $810.9 \div 100$.

2 Complete this pattern: 15, 11, 7, 3, ______

3 How many letters of the word MATHEMATICS are vowels? ______

4 What is the size of this angle?

5 Complete: 45.6 kg = ______ g

6 Evaluate $\sqrt{49}$.

7 List the first 5 multiples of 9.

8 Complete: $18 - 2 \times$ ______ $= 4$.

PART B: REVIEW

1 Evaluate $\frac{2}{3} - \frac{3}{5}$.

2 Find $\frac{1}{10}$ of \$8.

3 How many seconds in 4 minutes?

4 Convert to a simple fraction:

a 25%

b 10%

5 Find 50% of \$90.

6 Convert to a percentage:

a $\frac{3}{4}$

b 0.2

C S F

9780170454452

PART C: PRACTICE

› Percentages, fractions and decimals
› Percentage of a quantity
› Expressing quantities as fractions and percentages

1 Convert 66% to:

a a simple fraction

b a decimal

2 Find 72% of $250.

3 What percentage of 5 hours is 45 minutes?

4 Convert $12\frac{1}{2}\%$ to a simple fraction.

5 Convert to a percentage:

a $\frac{16}{40}$

b 0.015

6 12% of a school's students are left-handed. How many is this if the student population is 700?

PART D: NUMERACY AND LITERACY

1 Which is bigger: 30% or $\frac{1}{3}$? Show working.

2 Ziad scored 17 out of 20 in a science quiz. What is his mark as a percentage?

3 Complete: A percentage is a special fraction whose ________ is ________.

4 Jai bought a GPS navigator priced at $120 and received '15% off' this price.

a Calculate how much Jai saved.

b Calculate the sale price of the navigator.

5 How do you convert a fraction or decimal to a percentage?

6 Lisa works 40 hours per week and spends 6 of those hours travelling.

a What percentage of her time is spent travelling?

b What fraction of time is spent not travelling? Answer in simplest form.

HW HOMEWORK

C S F

4 FRACTIONS AND PERCENTAGES CROSSWORD

DO YOU KNOW WHAT AN 'IMPROPER FRACTION' IS?

PS PUZZLE SHEET

Across

1 To reduce a fraction to its most basic form

4 The line that separates the numerator from the denominator

5 A fraction with the numerator smaller than the denominator

9 Fractions that are the same

10 $\frac{1}{2}$ as a percentage

12 A fraction that is 'top-heavy'

14 A numeral that contains both a whole number and a fraction

16 $\frac{1}{5}$ as a percentage

17 The bottom number of a fraction

Down

2 What the symbol % stands for

3 Another name for 'one-fourth'

6 A percentage is a special fraction that has a denominator of one ____________.

7 The top number of a fraction

8 Multiplying by this fraction is the same as dividing by 2

11 A fraction with a denominator of 3

13 A fraction 'turned upside-down'

15 A number that shows part of a whole

18 Another word for simplify

9780170454452

STARTUP ASSIGNMENT 5

ALGEBRA AND EQUATIONS IS A NEW TOPIC, BUT IT'S BASED ON THINGS YOU ALREADY KNOW ABOUT NUMBER PATTERNS.

WS WORKSHEET

PART A: BASIC SKILLS

/ 15 marks

1 Simplify 15×100. ______

2 Draw a square and its axes of symmetry.

3 What number is 400 more than 7210?

______

4 Is 5.83 closer to 5.8 or 5.9? ______

5 True or false? $17 + 4 \leq 25 - 4$ ______

6 What angle is complementary to 70°? ______

7 Simplify $8 \times 4 \times 5$. ______

8 What type of angle is 158°? ______

9 What is the value of the 4 in 54 938?

______

10 Find $\frac{3}{5}$ of \$40. ______

11 Convert $\frac{2}{100}$ to a decimal. ______

12 Complete: 1 hour = ______ seconds

13 What is the measurement indicated by the arrow? ______

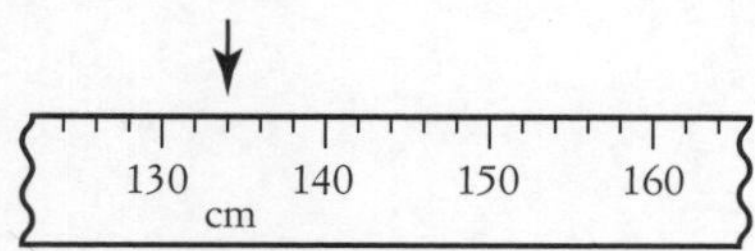

14 Simplify $\frac{12}{20}$. ______

15 Find the perimeter of this parallelogram.

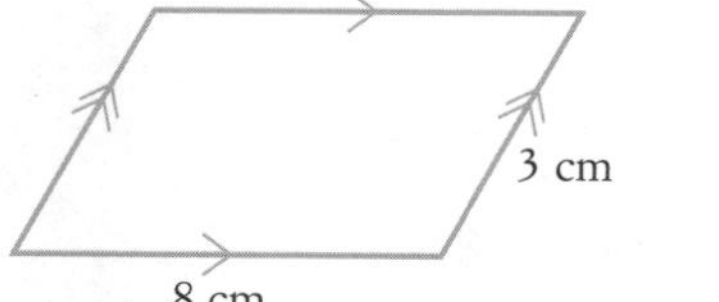

______

PART B: PATTERNS AND NUMBER

/ 25 marks

16 Find:

a $5 \times (-1)$ ______

b 2×7 ______

c $6 \times (-3) + 5$ ______

d $\frac{17 - 5}{4}$ ______

17 If $\Delta = 3$, what is the value of:

a $\Delta + 2$? ______

b $\Delta - 4$? ______

c $\Delta \times 4$? ______

18 Find the value of $\square$ if:

a $\square + 8 = 18$

______

b $\square - 4 = 5$

______

c $12 - \square = 9$

______

d $3 \times \square = 24$

______

9780170454452

WS WORKSHEET

19 Complete each number pattern.

a 3, 11, 19, 27, 35, ____________

b 16, 20, 24, 28, 32, ____________

c 18, 13, 8, 3, −2, ____________

20 Find:

a the product of 8 and 3 ____________

b the difference between 8 and 3 ____________

c the next odd number after 9 ____________

d 12 decreased by 5 ____________

21 **a** Complete the blanks in this number pattern.

3	×	1	+	4	=	7
3	×	2	+	4	=	__
3	×	3	+	4	=	__
__	×	__	+	__	=	__

b Look at the column of answers in part **a**.

What do you notice about them?

____

____

22 True or false? $14 \times (10 - 2) = 14 \times 10 - 14 \times 2$

____

____

23 If $\Delta = 7$, evaluate:

a $3 \times \Delta + 6$

____

b $\dfrac{\Delta + 3}{2}$

____

c $4 \times \Delta - 10$

____

d $(5 - \Delta) \times 2$

____

PART C: CHALLENGE

Bonus / 3 marks

In the sums below, each letter represents a digit.
Find the value of A, M, B and C.

```
  MA        AA
+  A      + BB
----      ----
  AM       CBC
```

C S F

9780170454452

SUBSTITUTION 5

WS WORKSHEET

1 If $r = 4$, then find the value of:

a $3 + r$ **b** $r - 5$ **c** $4r$ **d** $2r - 6$

e $r + 10$ **f** $10 - r$ **g** $3r + 1$ **h** $-2r + 2$

2 If $u = -6$, then evaluate:

a $u + 8$ **b** $2u$ **c** $u - 5$ **d** $5u + 12$

e $2 - u$ **f** $\frac{u}{-2}$ **g** $-u + 4$ **h** $2u - 1$

3 If $k = 2$ and $t = -2$, then evaluate:

a $k + t$ **b** kt **c** $3(k + 7)$ **d** $4t + 9k$

e $\frac{4k}{t}$ **f** $5kt$ **g** $3(k - t)$ **h** $k^2 + 5t$

4 Find the value of each expression.

a $3p - 2$ if $p = 5$ **b** $2m + 3$ if $m = -10$

c $15 - 3k$ if $k = 4$ **d** $2(6 - m)$ if $m = -4$

e $5p + 3$ if $p = 9$ **f** $m(m - 1)$ if $m = -2$

g $\frac{4 + m}{2}$ if $m = 20$ **h** $m^2 + m$ if $m = -3$

5 Use this table to evaluate each expression.

a	b	c	d	e	p	q	x	y	w
−3	5	7	2	4	−1	−8	10	1	100

a $a + b$ **b** $3a - c$ **c** $2b + y$ **d** $dw - x$

e p^2 **f** $2(c - q)$ **g** $w(c + q)$ **h** $qx - ac$

i $\frac{bd}{x}$ **j** $e^2 - ab$ **k** $by - q$ **l** $p^2 - q^2$

m $\frac{d + aw}{2}$ **n** $abcp$ **o** $9a + 3c - d$ **p** $bc - de$

q $\frac{dbw}{e}$ **r** $3w - 10pq$ **s** $8e - 2d - y$ **t** $\frac{d}{b} + \frac{y}{e}$

C S F

5 EQUATIONS MATCH

1 Enlarge this page onto cardboard for best results.

2 Cut out the 15 equations and solutions cards.

3 Match each equation to its correct solution.

EQUATIONS

1 $2x + 4 = 10$	**2** $2x - 6 = 3$	**3** $4x + 5 = 13$
4 $3x - 5 = 7$	**5** $\frac{x}{3} + 6 = 8$	**6** $\frac{6x}{5} = 18$
7 $\frac{x-1}{6} = 3$	**8** $\frac{3x}{8} = -3$	**9** $3x + 18 = 12$
10 $\frac{x+7}{2} = 9$	**11** $13 - 3x = -2$	**12** $11 - 4x = 23$
13 $3x + 8 = 35$	**14** $-3x + 4 = 22$	**15** $\frac{x}{4} - 7 = 3$

SOLUTIONS

A $x = 2$	**B** $x = 4$	**C** $x = 9$
D $x = 15$	**E** $x = 6$	**F** $x = 19$
G $x = 5$	**H** $x = 11$	**I** $x = -6$
J $x = -2$	**K** $x = 40$	**L** $x = 3$
M $x = 4\frac{1}{2}$	**N** $x = -8$	**O** $x = -3$

ALGEBRA FIND-A-WORD 

PUZZLE SHEET

N	G	U	X	L	A	R	E	M	U	N	O	R	P	E	J	T	A	U	N	A	L	L	Z	Q
O	Y	R	T	G	G	U	E	S	S	T	Z	W	F	E	P	A	M	I	G	M	V	C	N	H
I	L	F	O	R	M	U	L	A	S	X	V	J	E	K	B	L	A	N	R	M	C	G	H	B
S	G	I	U	X	B	Q	X	O	W	E	G	S	I	A	D	Q	R	Y	U	N	M	D	E	A
S	O	D	F	M	K	B	L	I	V	V	R	E	B	E	J	N	B	W	T	O	H	D	R	L
E	H	T	C	P	C	U	H	E	N	E	U	B	F	V	Q	I	E	A	O	I	R	I	P	A
R	U	D	K	O	T	P	U	F	V	L	R	D	S	A	T	J	G	L	M	T	W	S	A	N
P	E	U	K	I	I	I	H	N	A	E	K	C	G	L	J	S	L	F	F	A	N	T	I	C
X	J	D	O	F	R	G	I	V	V	V	U	L	F	U	Q	L	A	G	N	R	S	R	K	I
E	J	N	R	D	D	R	V	I	B	P	Z	C	P	A	R	I	T	H	M	E	T	I	C	N
I	B	K	Q	O	K	J	A	G	P	N	O	Z	S	T	Q	B	W	Z	X	P	D	B	O	G
L	M	M	W	X	C	T	O	E	E	N	T	U	B	E	A	R	U	E	I	O	D	U	P	S
Q	Q	P	P	S	I	K	V	H	S	L	B	A	N	C	G	O	Q	R	N	Z	O	T	P	D
E	V	T	R	O	O	L	Y	E	T	S	O	V	K	A	S	S	L	E	S	D	K	I	O	R
L	H	Q	N	O	O	J	C	L	T	Y	S	T	R	O	X	V	H	S	E	H	F	V	S	L
B	N	U	F	S	V	U	F	I	K	P	R	E	X	P	A	N	D	E	D	M	B	E	I	J
A	S	C	M	J	T	E	T	P	E	A	A	S	S	O	C	I	A	T	I	V	E	U	T	R
I	Y	Z	Y	I	A	U	R	T	C	P	T	I	O	O	W	Y	Z	V	I	P	M	E	E	E
R	V	W	V	G	T	O	U	K	X	E	S	T	E	P	J	T	A	U	N	A	L	L	Z	V
A	Q	E	Y	I	B	R	I	T	G	T	Z	W	C	O	M	M	U	T	A	T	I	V	E	E
V	F	E	O	L	P	N	A	M	I	G	M	V	C	N	H	L	X	V	J	K	B	L	N	R
R	M	N	E	C	G	G	H	G	I	U	X	B	Q	X	W	E	G	I	D	Q	Y	U	M	S
D	E	M	O	D	F	M	K	B	I	V	V	J	N	T	H	R	H	T	C	M	R	E	T	E
P	C	H	E	K	C	E	H	C	N	F	Q	I	O	R	P	U	D	K	O	P	U	F	D	S
T	J	M	W	A	U	K	I	I	H	K	C	G	J	S	F	F	N	O	I	T	A	U	Q	E

Find these words in the puzzle above. They are across, up and down, and diagonal, and can be backwards as well as forwards.

ABBREVIATION	CONSECUTIVE	IMPROVE	REVERSE
ALGEBRA	DISTRIBUTIVE	INVERSE	SOLVE
ARITHMETIC	EQUATION	LAW	SOLUTION
ASSOCIATIVE	EXPANDED	OPERATION	STEP
BACKTRACKING	EXPRESSION	OPPOSITE	SUBSTITUTION
BALANCING	EVALUATE	ORDER	TERM
CHECK	FORMULA	PROBLEM	VALUE
COMMUTATIVE	GUESS	PRONUMERAL	VARIABLE

5 MENTAL CALCULATION

Name:

Due date:

Parent's signature:

Part A	/ 8 marks
Part B	/ 8 marks
Part C	/ 8 marks
Part D	/ 8 marks
Total	/ 32 marks

HW HOMEWORK

PART A: MENTAL MATHS

Calculators are not allowed.

1 Complete: 3838 is changed to 3238 by subtracting ____________.

2 Simplify $\frac{14}{20}$. ____________

3 Find the perimeter of this triangle.

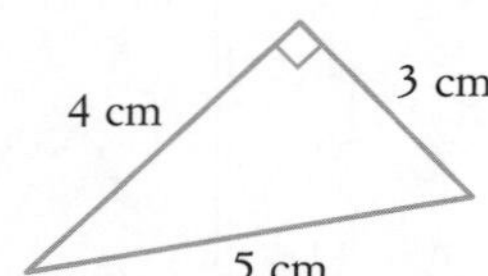

4 List these integers in ascending order:

0, −7, 5, −3, 6 ____________

5 Complete: 4.65 L = ____________ mL

6 Draw an obtuse angle.

7 List the factors of 12. ____________

8 Find the highest common factor of 12 and 20.

____

PART B: REVIEW

1 If □ = 4, evaluate 2 × □ + 7.

2 Evaluate each expression.

a $3 \times (20 + 2)$

____

b $3 \times 20 + 3 \times 2$

____

3 True or false?

a $18 + 2 = 2 + 18$ ____________

b $18 - 2 = 2 - 18$ ____________

c $18 \times 2 = 2 \times 18$ ____________

d $18 \div 2 = 2 \div 18$ ____________

4 What number is 1 less than double 6?

____

C S F

9780170454452

PART C: PRACTICE

› Mental calculation
› The laws of arithmetic

1 Evaluate each expression by mental calculation.

a 900×4

____

____

b $-7 + 18$

____

____

c $12 + 6 + 19 + 31 + 24$

____

____

d $3 \times 6 \times 5$

____

____

2 Show the distributive law in each multiplication.

a $25 \times 8 = 25 \times (10 - 2)$

$=$ ______________________

$= 200$

b $17 \times 11 = 17 \times (10 + 1)$

$=$ ______________________

$= 187$

3 For any 3 numbers a, b and c, is each equation true or false?

a $(a + b) + c = a + (b + c)$

____

b $(a - b) - c = a - (b - c)$

____

PART D: NUMERACY AND LITERACY

1 Evaluate each expression by mental calculation and explain how you found the answer.

a 3×80

____

____

____

b $7 \times 5 \times 4$

____

____

____

c $63 + 18 + 12 + 7$

____

____

____

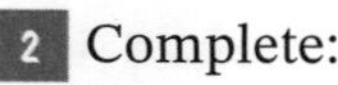

2 Complete:

a The rule $a \times b = b \times a$ is called

the ______________________

law of ______________________.

b The distributive law is:

$a(b + c) =$ ______________________.

3 (2 marks) Use the distributive law to evaluate 37×9 and explain how you found the answer.

____

____

____

5 ALGEBRA

Name:

Due date:

Parent's signature:

Part A	/ 8 marks
Part B	/ 8 marks
Part C	/ 8 marks
Part D	/ 8 marks
Total	/ 32 marks

PART A: MENTAL MATHS

Calculators are not allowed.

1 Complete: $5 \times 8 = 2 \times$ ______

2 How many degrees in a revolution?

______

3 Complete this number pattern:

5, 10, 20, 40, ______

4 Find the average of 9 and 17.

______

5 Complete: 5 minutes = ______ seconds

______

6 Draw the axes of symmetry on this rectangle.

7 Find the sum of the first 3 prime numbers.

______

8 Evaluate $174 \div 3$.

______

PART B: REVIEW

1 Evaluate each expression by mental calculation.

a $4 \times 8 \times 25 \times 2$

______

b 60×30

______

c 35×11

______

2 Complete each equation.

a ______ $+ 9 = 14$

b $3 \times$ ______ $= 27$

c ______ $\div 5 = 10$

d $2 \times$ ______ $- 4 = 8$

3 Complete the distributive law:

$a(b - c) =$ ______

C S F

HW HOMEWORK

PART C: PRACTICE

› Algebraic expressions
› Substitution

1 Simplify each expression.

a $3 \times a \times b$ ______

b $r + 2r$ ______

c $9 - d \times 3$ ______

2 Write each statement as an algebraic expression using n to stand for the number.

a The sum of 4 and the number

______

b 3 times the number, then decrease by 6

______

3 If $a = 3$, evaluate each expression.

a $3a + 4$

______

b $a^2 - 2a$

______

c $8 - a$

______

2 Write the general rule for this pattern, using a variable:

$-4 \times (-4) = 4^2$

$-3 \times (-3) = 3^2$

$-10 \times (-10) = 10^2$

$-7 \times (-7) = 7^2$

______

3 Write in words the meaning of each algebraic expression.

a $4w^2$

______

______

b $\dfrac{y+5}{2}$

______

______

4 Explain what $n \div n$ simplifies to, giving reasons.

______

______

5 What word means to replace a variable with a value in an algebraic expression or formula?

______

PART D: NUMERACY AND LITERACY

1 Write an algebraic expression for each statement.

a The change from \$20 after spending \$$p$

______

b How many times n divides into 18

______

6 The formula for converting Celsius temperatures to Fahrenheit temperatures is $F = \frac{9}{5}C + 32$. Use it to convert 25°C to Fahrenheit.

______

______

HW HOMEWORK

C S F

5 EQUATIONS 1

Name:

Due date:

Parent's signature:

Part A	/ 8 marks
Part B	/ 8 marks
Part C	/ 8 marks
Part D	/ 8 marks
Total	/ 32 marks

WORK THROUGH THESE PROBLEMS TO BECOME A MASTER AT ALGEBRA AND SOLVING EQUATIONS.

HW HOMEWORK

PART A: MENTAL MATHS

Calculators are not allowed.

1 Complete: $3 \times 8 =$ ________ $\times 6$

2 What does 13:50 in 24-hour time mean?

____

3 List the first 5 multiples of 4.

____

4 How many degrees in a straight angle?

____

5 Find 25% of \$92.

____

____

6 Complete: 674 mg = ________ g

7 Evaluate 16×9.

____

8 A rectangle 9 cm long has a perimeter of 26 cm. What is its height?

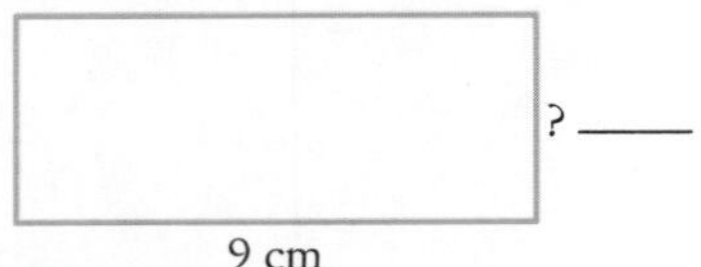

PART B: REVIEW

1 Simplify each expression.

a $b + b - b$

____

b $2 \times w \times w$

____

c $10 \div (d \times e)$

____

2 Write an algebraic expression for:

a the difference between x and 3 times y

____

b the square root of double t

____

c the average of a and b

____

3 Evaluate $2m - 5$ if:

a $m = 4$

____

b $m = -4$

____

C S F

PART C: PRACTICE

› One-step equations
› Two-step equations

Solve each equation.

1 $4q = 30$

2 $\frac{d}{8} = -5$

3 $z + 9 = 8$

4 $r - 3 = -3$

5 $2n - 4 = 12$

6 $3e + 2 = 11$

7 $\frac{n}{4} - 6 = 5$

8 $\frac{y + 3}{9} = 2$

C
S
F

PART D: NUMERACY AND LITERACY

1 (2 marks) Solve $8n - 14 = 42$ by 'guess, check and improve'. Show all working.

2 What is the name given to a letter such as x in an algebraic expression?

3 **a** Describe what the equation $5c - 4 = 11$ means in words.

b (2 marks) What are **inverse operations?** Give an example of inverse operations.

c Solve $5c - 4 = 11$, showing the inverse operations clearly.

d Check that your solution is correct.

HW HOMEWORK

5 EQUATIONS 2

Name:

Due date:

Parent's signature:

Part A	/ 8 marks
Part B	/ 8 marks
Part C	/ 8 marks
Part D	/ 8 marks
Total	/ 32 marks

PART A: MENTAL MATHS

Calculators are not allowed.

1 Evaluate $9 - (-9)$. ______

2 What is an isosceles triangle?

______

______

3 Mark $\angle DEC$.

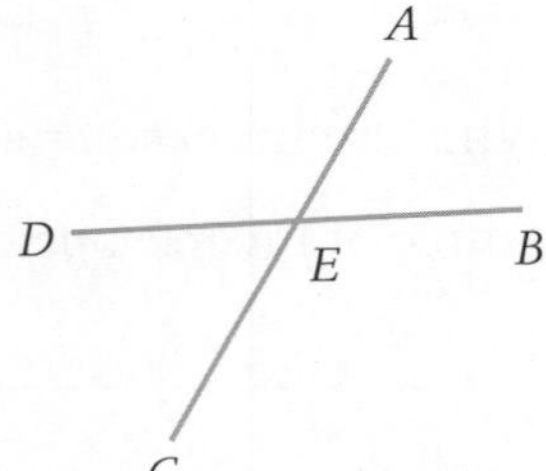

4 How many faces does a cube have? ______

5 Complete: 72 hours = ______ days

6 List the first 5 multiples of 8.

______

7 Find the lowest common multiple (LCM) of 5 and 2. ______

8 Complete: If a number is divisible by 6, it must be divisible by both ______ and ______.

PART B: REVIEW

1 Solve each equation, showing all steps.

a (2 marks) $3a - 4 = 8$

______

______

______

b (2 marks) $\frac{y + 3}{5} = 7$

______

______

______

2 This rectangle has length l and width w.

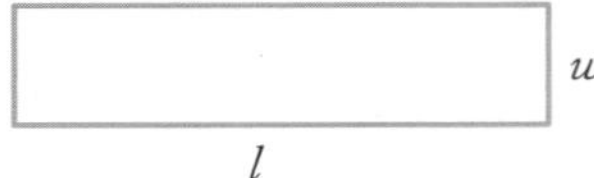

Write a simplified algebraic expression for its:

a perimeter ______

b area ______

3 Write an algebraic expression for:

a the sum of d and the square root of e

______

b the cost of one movie ticket if 4 cost $\$p$

______

C S F

9780170454452

PART C: PRACTICE

› Substitution
› Equation problems

1 (4 marks) Write an equation for each problem, then solve the equation.

a The difference between a number and 14 is 8. What is the number?

______

______

______

b Ned's age doubled, less 6 is equal to 18. How old is Ned?

______

______

______

2 Find the value of $\frac{x + 6}{2}$ if:

a $x = 10$ ______

b $x = -6$ ______

3 The area of a triangle has the formula $A = \frac{1}{2}bh$ where b is the length of the base and h is the perpendicular height.

a Find the area of a triangle with base 8 cm and height 18 cm.

______

______

b Find the height of the triangle with base 10 cm and area 35 cm^2.

______

______

PART D: NUMERACY AND LITERACY

1 Write the general rule for this pattern, using a variable:

$6 + (-6) = 0$

$2 + (-2) = 0$

$9 + (-9) = 0$

$12 + (-12) = 0$

______

2 (2 marks) Solve $38 - 3k = 11$ by guessing and checking. Show working.

______

______

______

3 Write the word that means:

a the answer to an equation

______

b a formal rule written as an algebraic equation

______

4 Explain why $1x = x$.

______

______

5 Write algebraically the commutative law for multiplication: any 2 numbers can be multiplied in any order.

______

6 Write an algebraic expression for the change from \$$n$ after buying y theme park admission tickets for \$25 each.

______

HW HOMEWORK

WS WORKSHEET

6 STARTUP ASSIGNMENT 6

THE GEOMETRICAL FIGURES TOPIC COVERS TRANSFORMATIONS, SYMMETRY, TRIANGLES AND QUADRILATERALS, SO THERE ARE SOME RULES, SYMBOLS AND TERMINOLOGY THAT YOU NEED TO KNOW.

PART A: BASIC SKILLS

/ 15 marks

1 Round 2107 to the nearest ten. ______

2 Evaluate:

a $-4 - 10$ ______

b $4 + (-10)$ ______

3 Complete: 0.8 kg = ______ g

4 **a** Describe what a prime number is.

______

______

b Write a prime number between 10 and 20.

______

5 **a** What is the name of this solid?

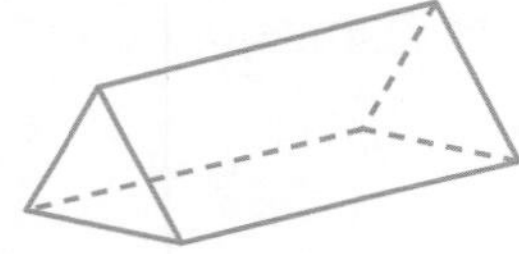

______

b How many faces has this solid? ______

6 Use a factor tree to write 150 as a product of its prime factors.

______

7 Convert 5% to:

a a decimal ______

b a simple fraction ______

8 Evaluate:

a 18×200 ______

b $845 \div 5$ ______

9 If $p = 7$, then evaluate $4p + 2$. ______

10 What is another name for a 180° angle?

______

PART B: ANGLES AND SHAPES

/ 25 marks

11 **a** Mark a pair of cointerior angles on this diagram.

b What is the property of cointerior angles on parallel lines?

______

______

12 **a** Draw a rectangle.

b Draw its axes of symmetry.

13 **a** Draw an equilateral triangle.

b Does it have rotational symmetry? ______

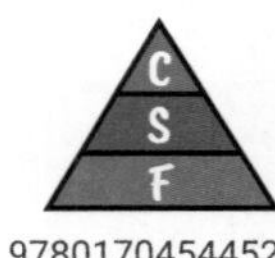

9780170454452

14 Write the symbol for:

a 'is parallel to' ______________________

b 'is perpendicular to' ______________________

15 **a** Measure this angle.

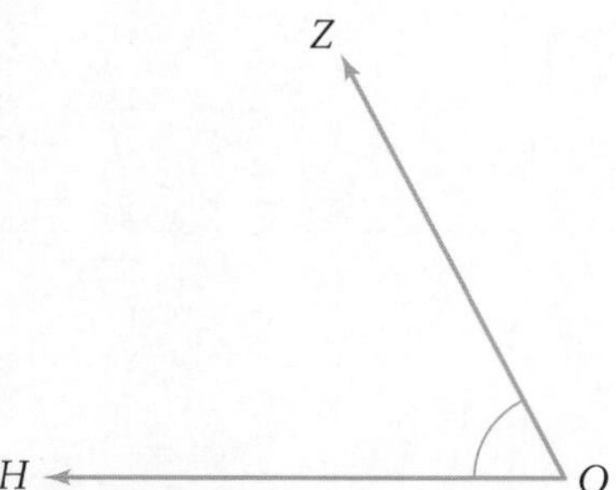

b State what type of angle it is.

____

c Name the angle using 3 letters. ______________________

16 How many sides has a hexagon? ______________________

17 Does rotation mean a flip, slide or spin?

____

18 What angle size is complementary to 50°?

____

19 Name each shape.

a

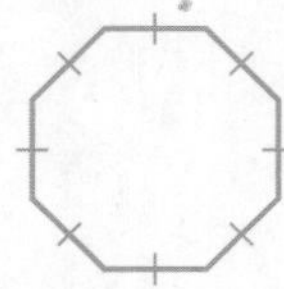

b

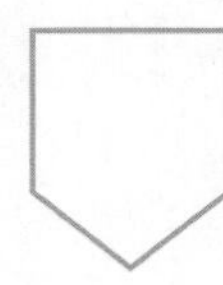

______________________ ______________________

20 Draw:

a a trapezium

b a right-angled triangle

21 Find y.

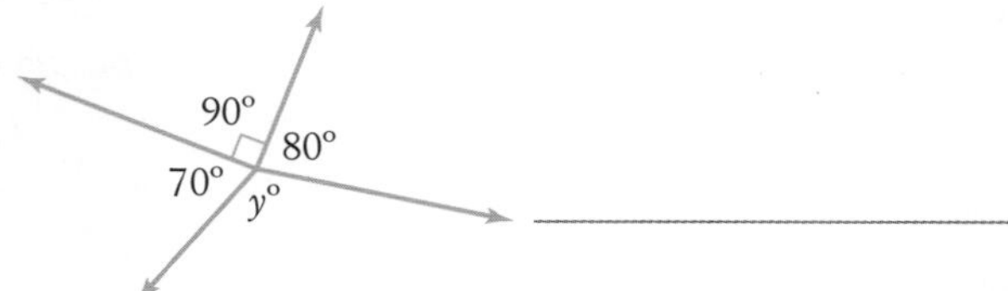

____

22 **a** What is a quadrilateral?

____

____

b Name an example of a quadrilateral.

____

23 How many axes of symmetry has:

a a kite?

b a parallelogram?

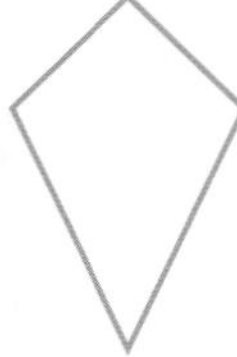

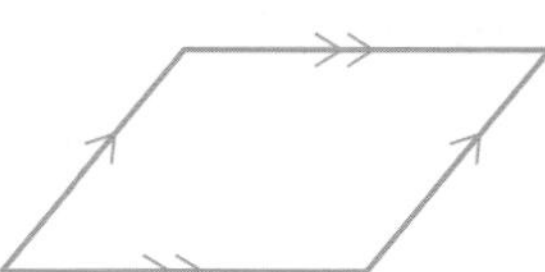

______________________ ______________________

24 **a** What is an obtuse angle?

____

____

b Draw an obtuse angle.

PART C: CHALLENGE

Bonus / 3 marks

How many squares can be found on this chessboard? (The answer is a lot more than 64)

C
S
F

6 TRANSFORMATIONS

THERE ARE 3 GEOMETRICAL TRANSFORMATIONS OR MOVEMENTS ON SHAPES. ONLY THE POSITION CHANGES, NOT THE SIZE OR SHAPE.

WS WORKSHEET

1 Translation, reflection or rotation?

a To make a shape back-to-front to create a mirror-image ______

b To slide or shift a shape ______

c To flip a shape ______

d To turn a shape upside-down ______

e To move a shape up, down or diagonally ______

f To spin a shape ______

g 

______

h F →

______

i K →

______

j R →

______

k S →

______

l W →

______

2 Translate each shape according to the directions given.

a

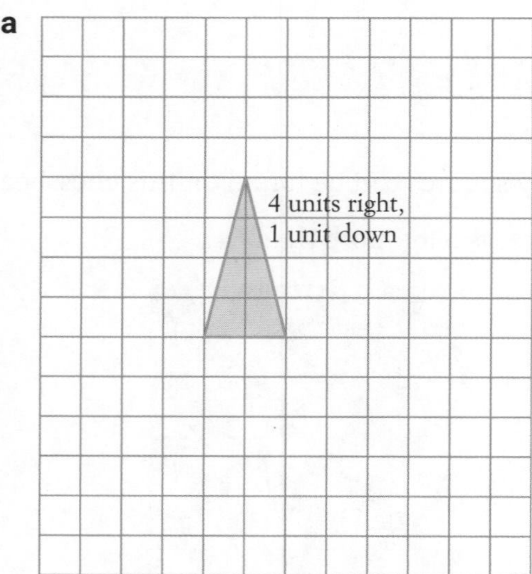

b

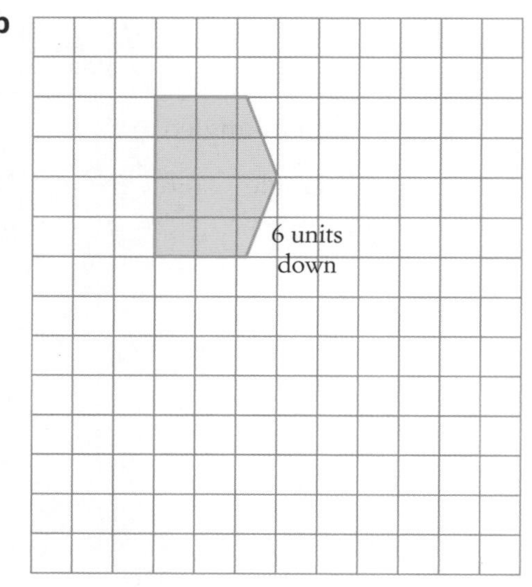

9780170454452

c

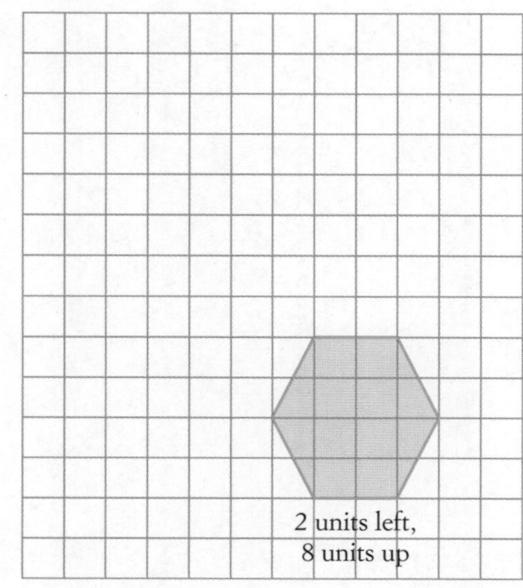

d

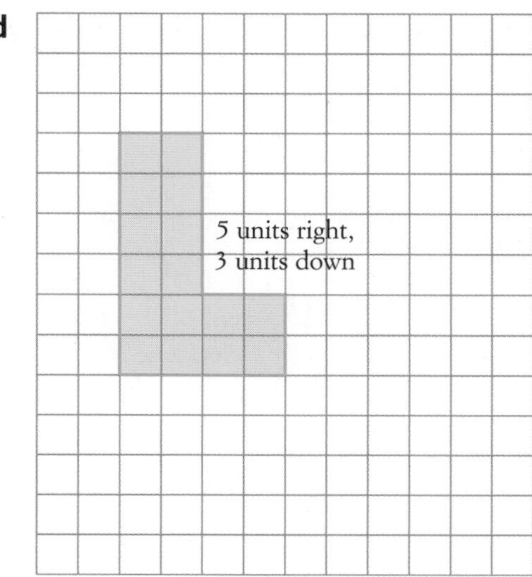

 Reflect each shape across the dotted line. Matching points on the original and reflected shape should be the same distance from the line of reflection.

a

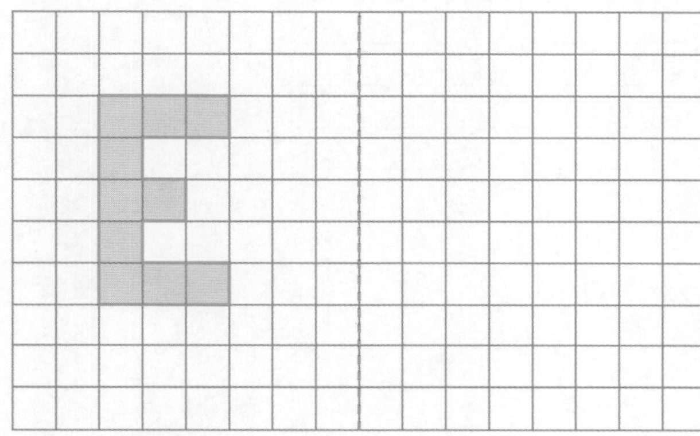

b

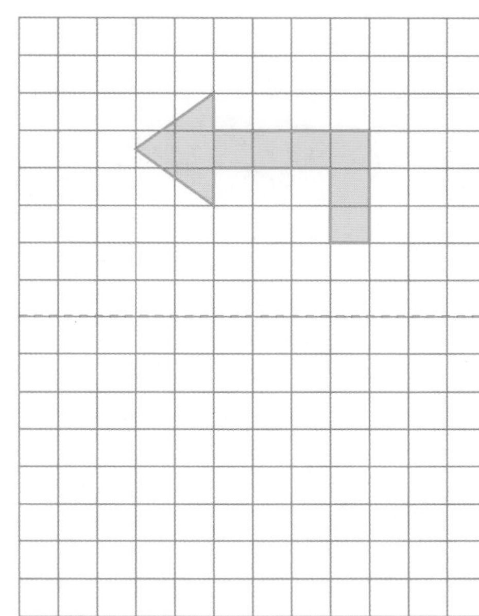

c

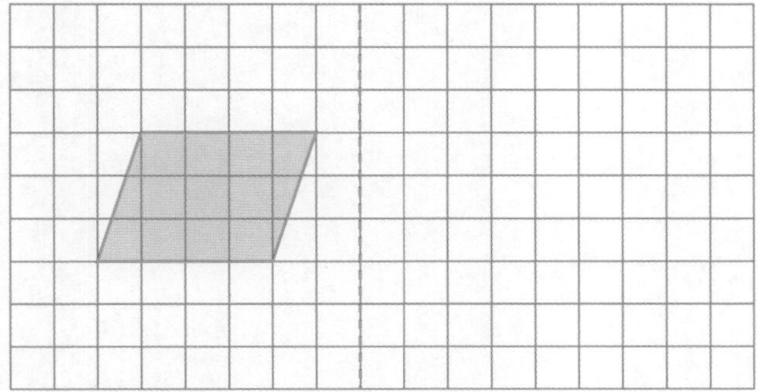

d

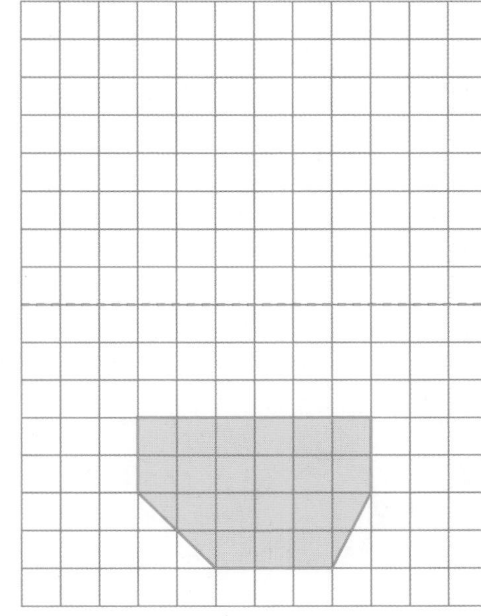

WS WORKSHEET

C S F

WS WORKSHEET

4 Rotate each shape by the given angle about the point O.

a

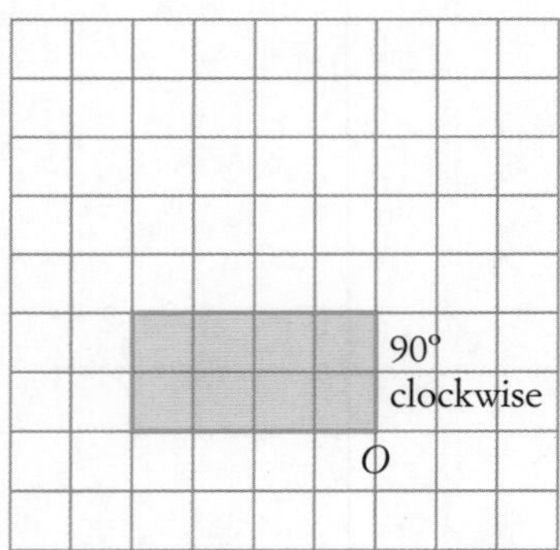

b

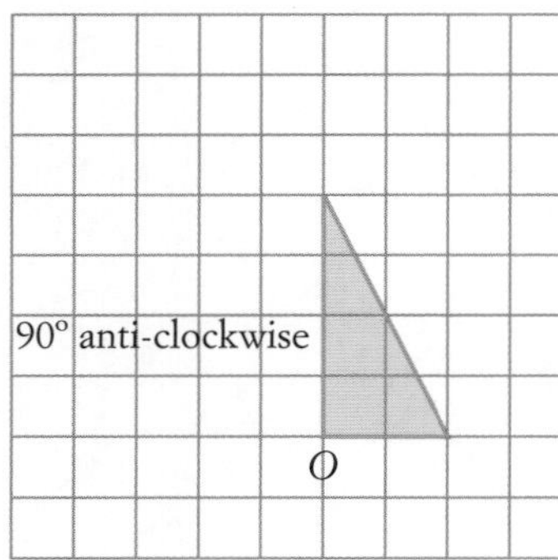

c

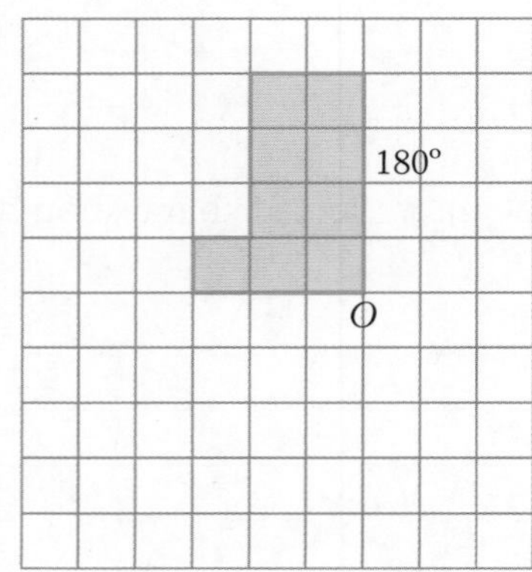

d

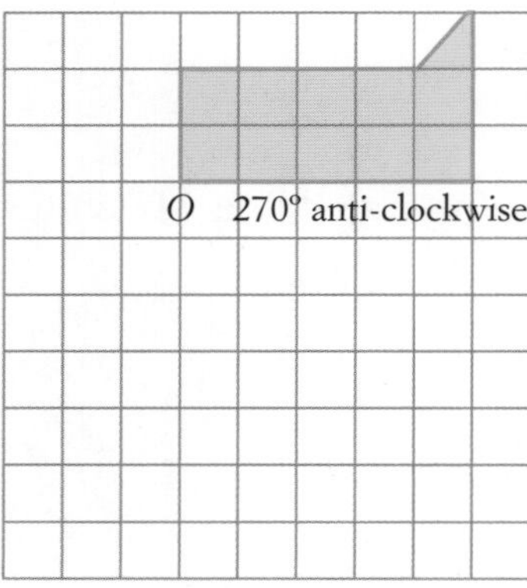

C S F

9780170454452

For each shape:

- draw its axes of symmetry
- state the order of rotational symmetry
- count its axes of symmetry
- mark its centre of symmetry.

1 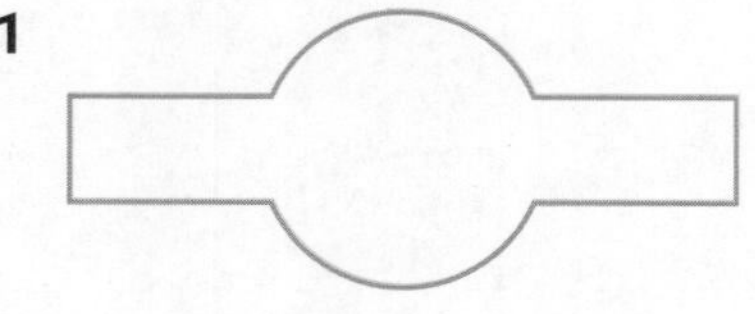

2 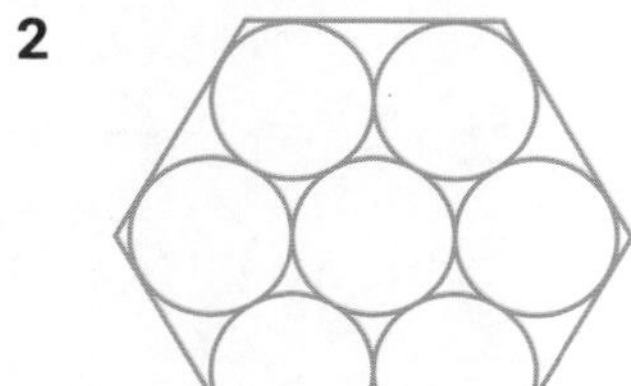

3 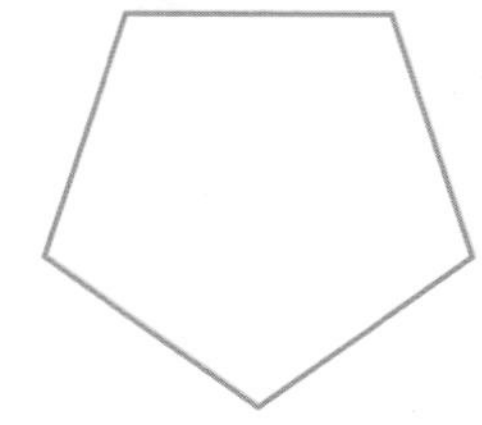

4 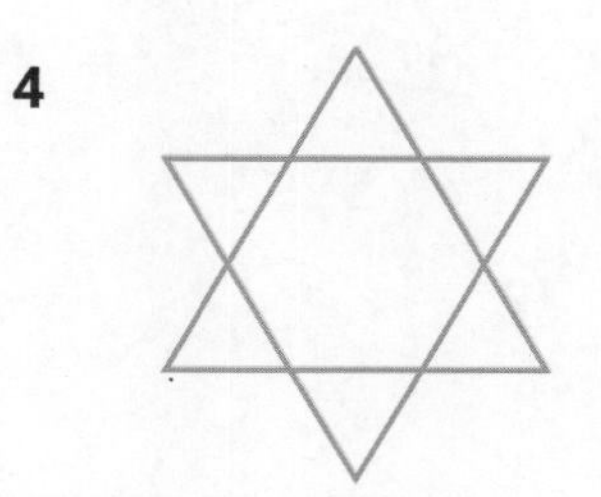

5 

6 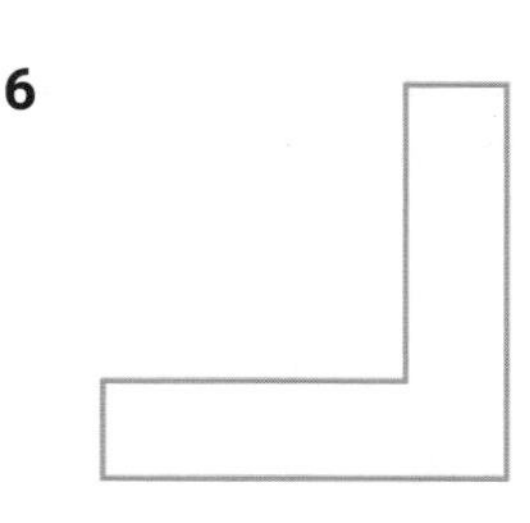

7 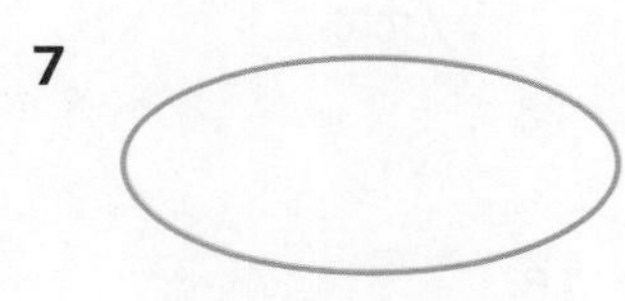

8 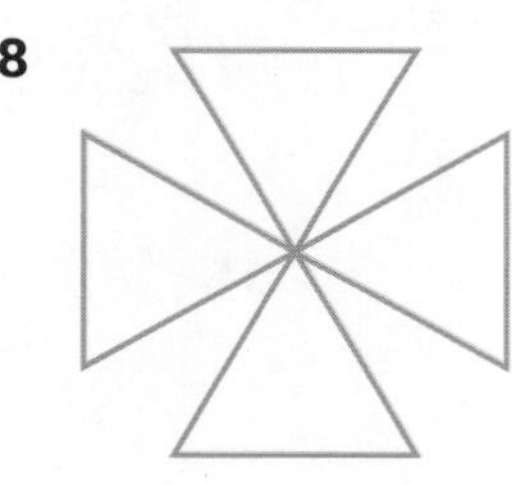

9 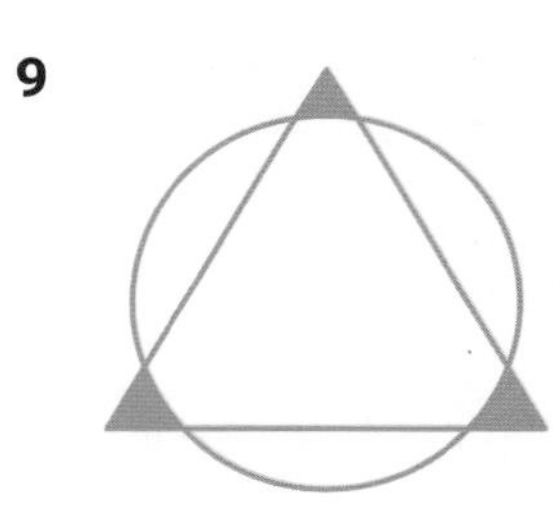

10 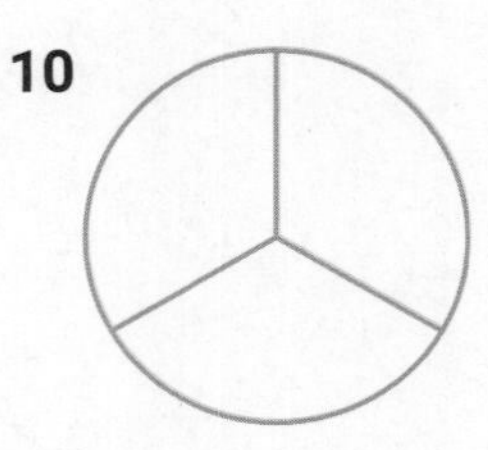

11 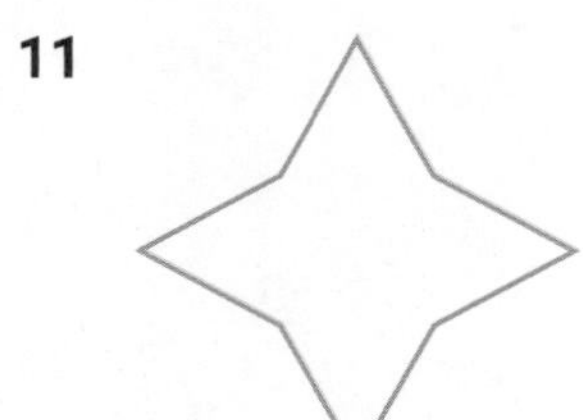

12 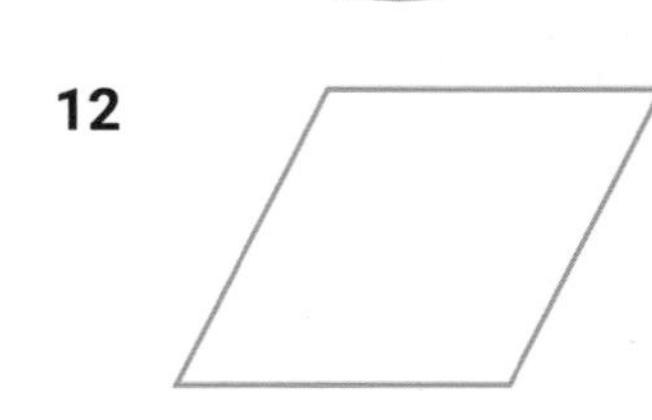

13 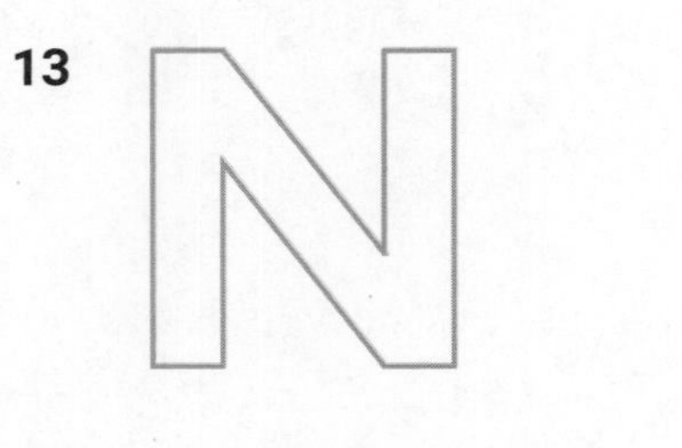

14 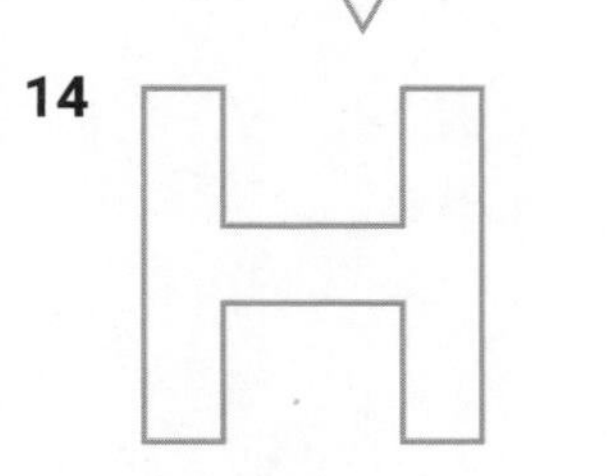

15 

16 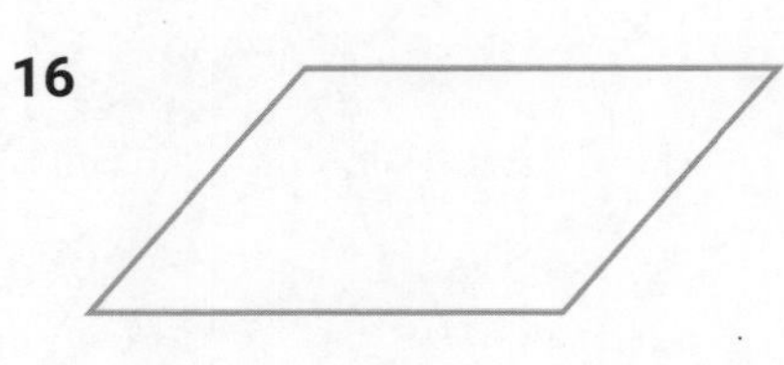

17 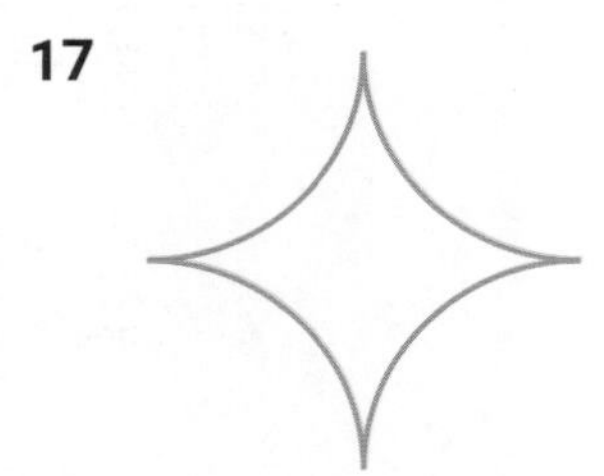

18 

6 TRIANGLE GEOMETRY

USE YOUR GEOMETRY SKILLS TO FIND THE MISSING ANGLES.

- Angle sum of a triangle
- Equilateral and isosceles triangles
- Exterior angle of a triangle

Find the value of the variable(s) in each diagram.

1
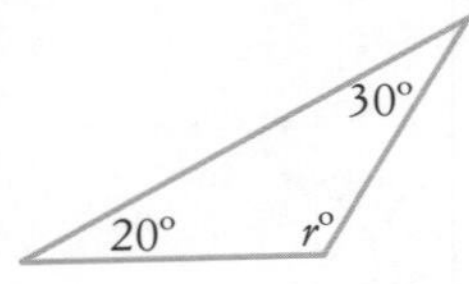

2
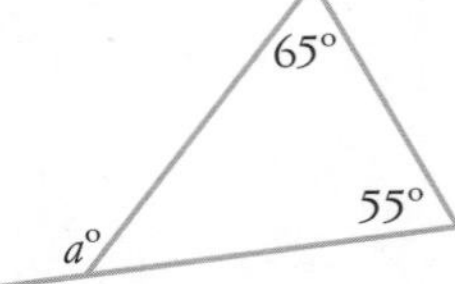

3
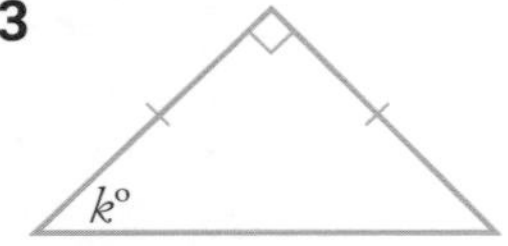

4
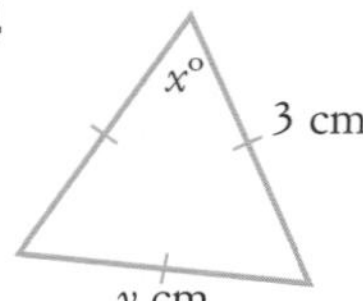

5
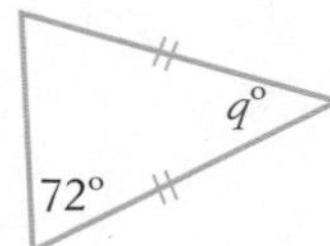

6
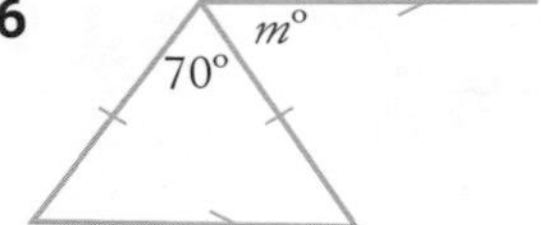

7
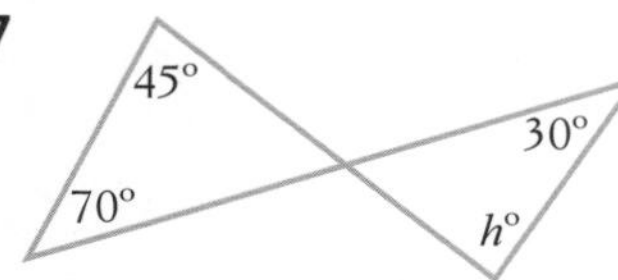

8
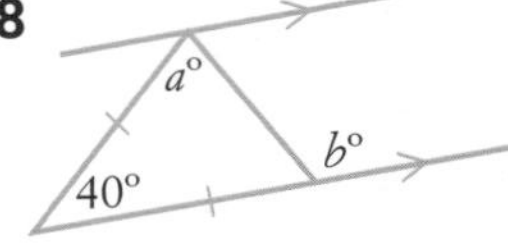

9
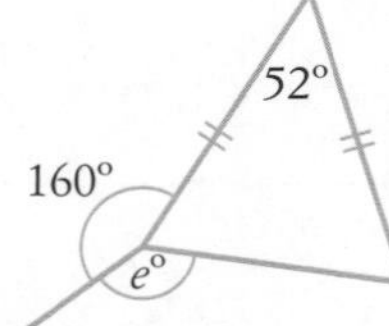

10
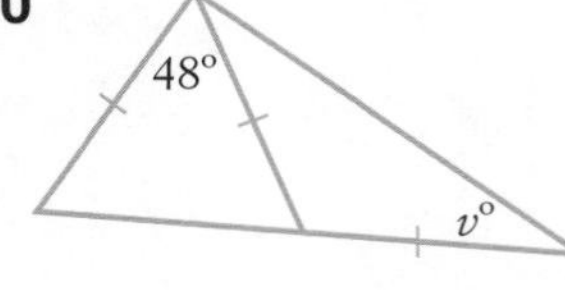

11
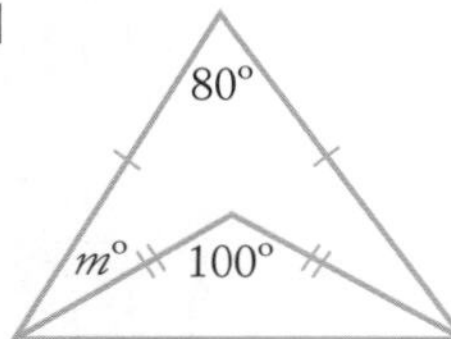

12
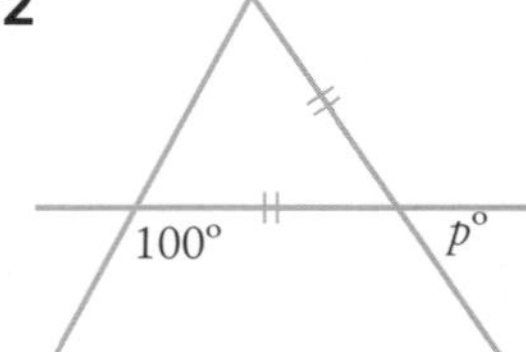

13
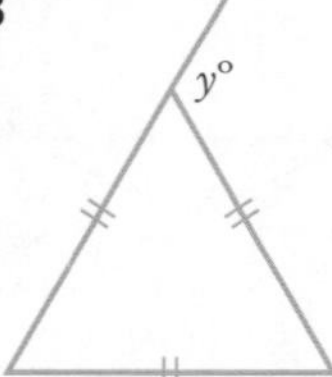

14
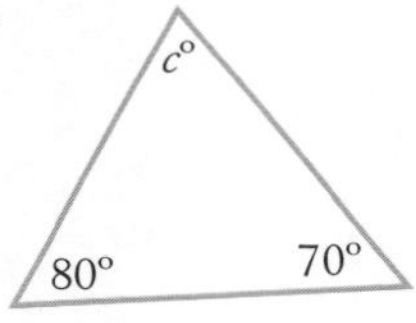

15
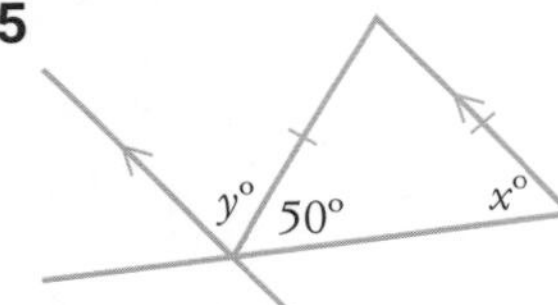

16
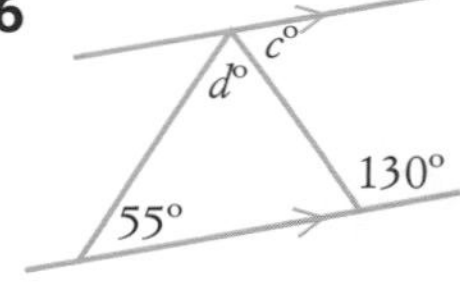

17
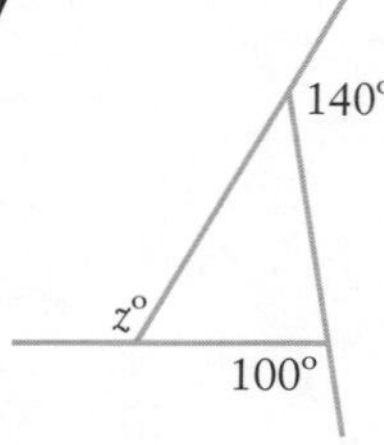

18
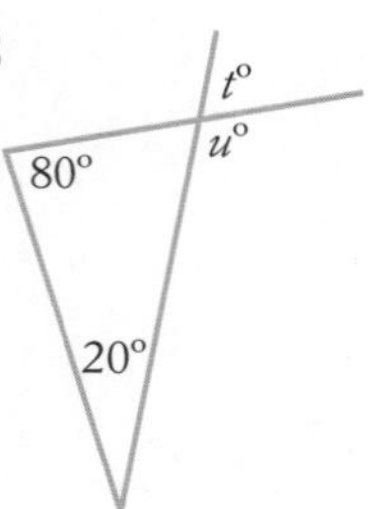

19
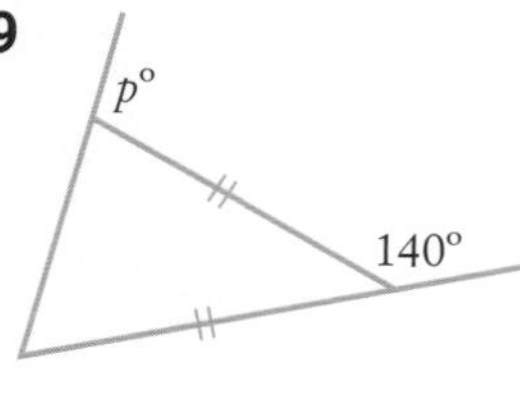

20
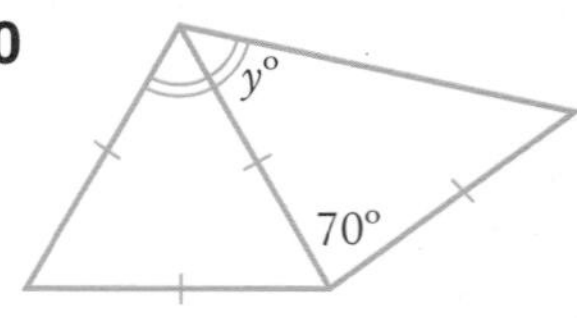

21
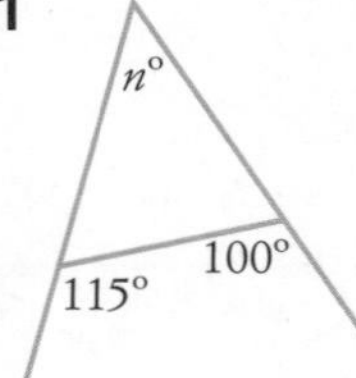

22
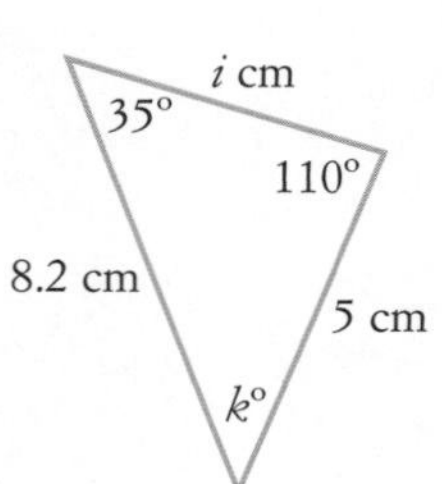

23
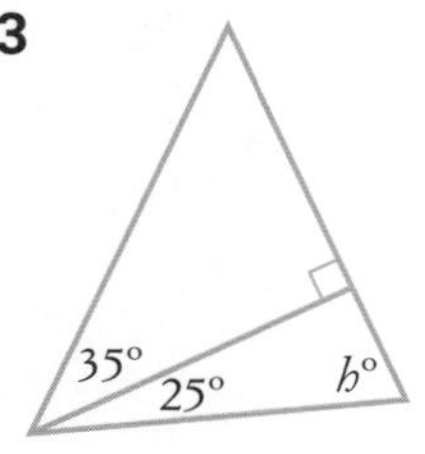

24
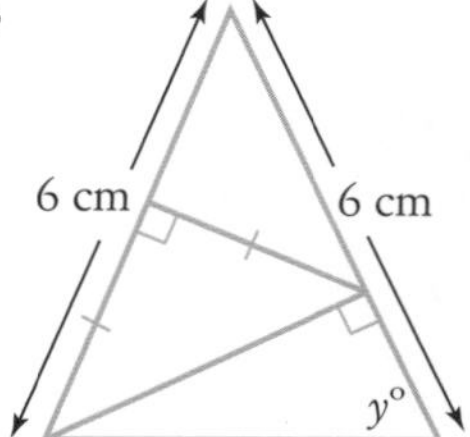

9780170454452

GEOMETRY FIND-A-WORD 6

PS PUZZLE SHEET

E E F E G C A D J A C E N T L A N O I T A T O R Y

L U U S O S N G N O G A C E D A M U I Z E P A R T

L I L G F T N O I T C E L F E R T C A R T O R P L

I P P E N T A G O N J W L A N O G A I D E C T F X

P H D A T C E S I B H O E T I K P R E L U D K V S

S S U E N I L E X T E R I O R E D U L C N I R U W

E N E L A C S I S O S C E L E S I W N O G Y L O P

X T H G I R K Q U A D R I L A T E R A L G A N O N

A X E S I N O G A C E D O D G E I M E T R I C A L

W C R A N O S I X A G H E P T M G O N E S U T B O

M E L E D N R M A I D O Q S A E T I S O P P O D Y

S E T T R A O U G E R N O G A T C O S U B M O H R

E Q U I L A T E R A L R E S Q U A R E N A L P V T

E A N S O H A R A L U C I D N E P R E P E N P L E

Y G O O Q U T L E Y F I S S A L C S I D E I J Y M

G X G P Y U I L E L G N A T C E R G E T U C A I M

I I A M S T O F F U T C U R T S N O C T D B N I Y

V X X O U R N K M A R G O L E L L A R A P O N B S

R H E C I T R A N S L A T I O N R Q W S T N I O P

U M H C L O F W D E L G N A V L J Q E R U G I F A

Find these words in the puzzle above. They are across, up and down, and diagonal, and can be backwards as well as forwards.

ACUTE	ANGLED	AXES	AXIS
CLASSIFY	COMPOSITE	EQUAL	EQUILATERAL
EXTERIOR	IMAGE	ISOSCELES	KITE
LINE	OBTUSE	OPPOSITE	ORDER
PARALLELOGRAM	PERPENDICULAR	POINT	QUADRILATERAL
RECTANGLE	REFLECTION	RHOMBUS	RIGHT
ROTATION	ROTATIONAL	SCALENE	SIDE
SQUARE	SYMMETRY	TRANSLATION	TRAPEZIUM

6 TRANSFORMATIONS AND SYMMETRY

YOU'LL NEED YOUR SEEING AND DRAWING SKILLS FOR THIS ASSIGNMENT AS THERE IS A LOT OF DIAGRAM WORK INVOLVING SHAPES. HAVE A RULER, PENCIL AND YOUR EYES READY.

Name:

Due date:

Parent's signature:

Part A	/ 8 marks
Part B	/ 8 marks
Part C	/ 8 marks
Part D	/ 8 marks
Total	/ 32 marks

HW HOMEWORK

PART A: MENTAL MATHS

Calculators not allowed.

1 Convert $\frac{1}{4}$ to a percentage.

2 Evaluate $-2 - (-3) + (-1)$.

3 Write an algebraic rule relating a and b.

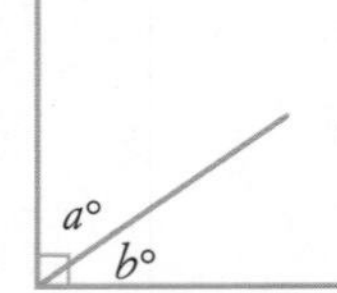

4 Convert 0.8 to a simple fraction.

5 If $x = 4$ and $y = 6$, evaluate $x^2 + 3y$.

6 What are the possible outcomes when a coin is tossed?

7 Solve the equation $5m - 12 = 23$.

8 Use a factor tree to write 54 as a product of its prime factors.

PART B: REVIEW

1 Name the 3 types of transformations.

2 Translate the L-shape 5 units right and 3 units down.

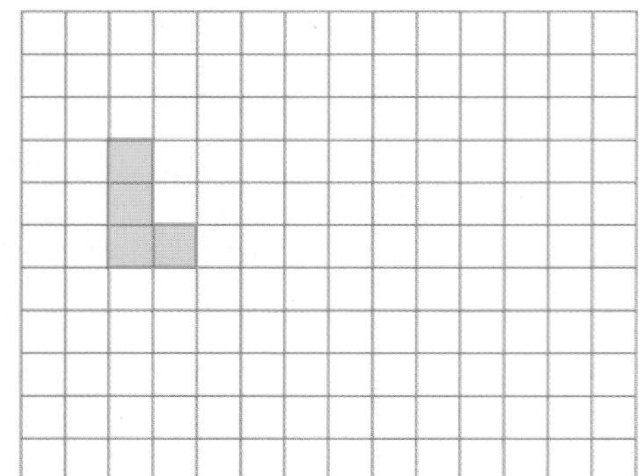

C S F

9780170454452

3 How many axes of symmetry has:

a a square?

____

b a regular pentagon?

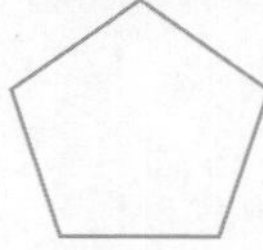

____

c an ellipse?

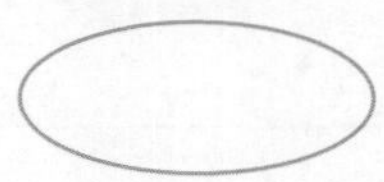

____

4 Rotate the L-shape 180° about P.

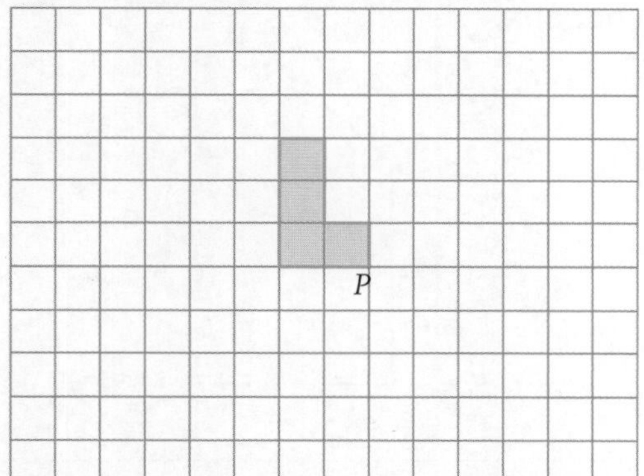

5 Which transformation has been performed on the letter R?

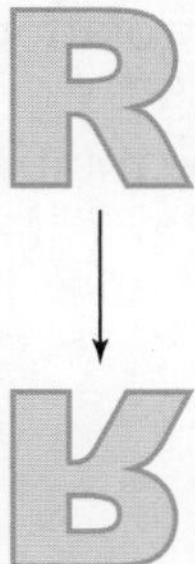

6 Reflect the L-shape across the line.

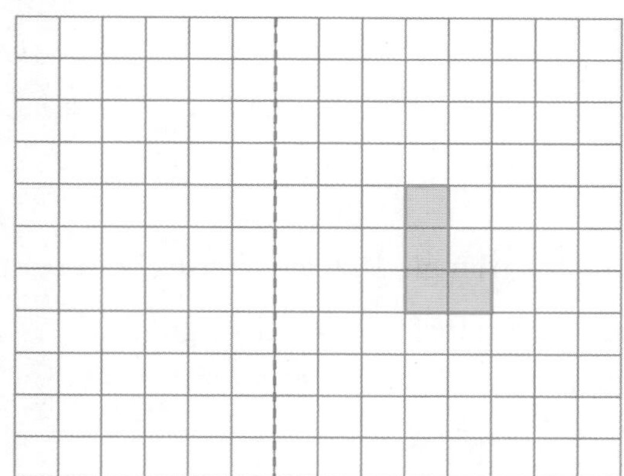

PART C: PRACTICE

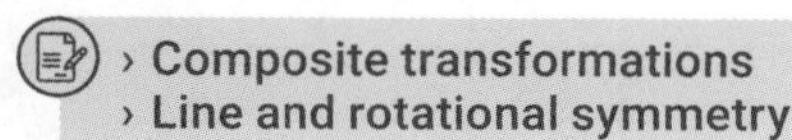

1 Reflect the triangle across the line, then translate it 5 units down, 2 units left.

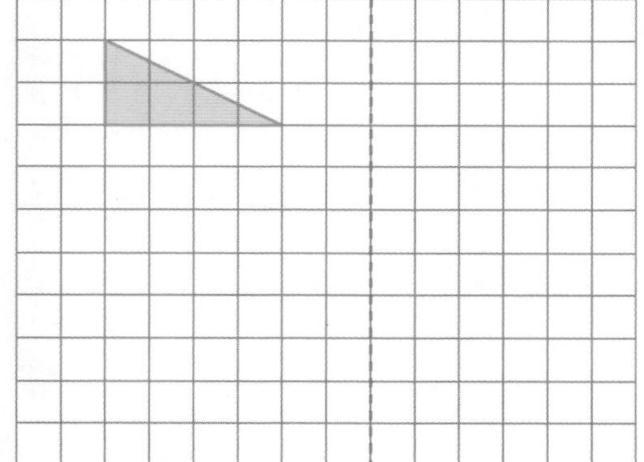

2 For each shape, draw all axes of symmetry.

a

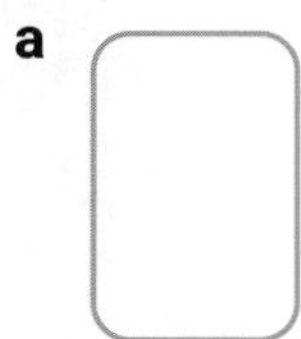

b

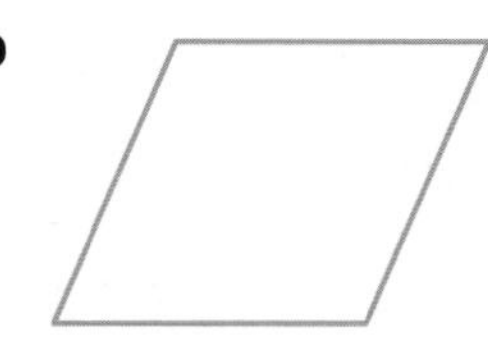

3 Draw a shape that has one axis of symmetry.

4 Explain why a circle has an infinite number of axes of symmetry.

5 Rotate the triangle 180° clockwise about point A and then translate it 4 units right

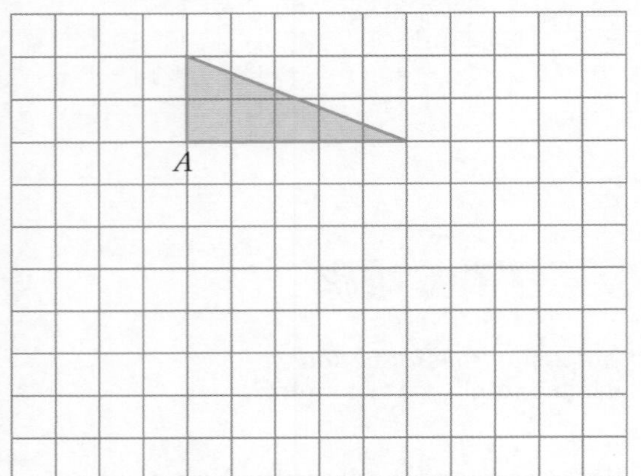

6 For each polygon, write its order of rotational symmetry.

a

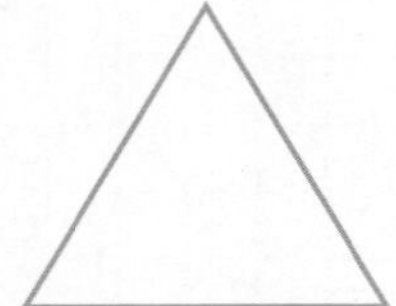

b

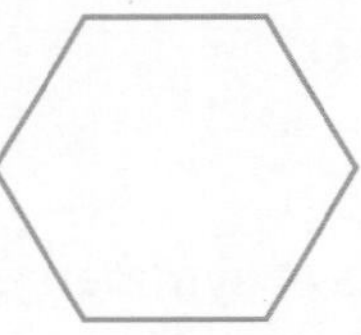

PART D: NUMERACY AND LITERACY

1 What name is given to a shape after it has been transformed?

2 Which transformation involves 'flipping' a shape?

3 Describe what rotation means.

4 Which transformation has been made on Shape A:

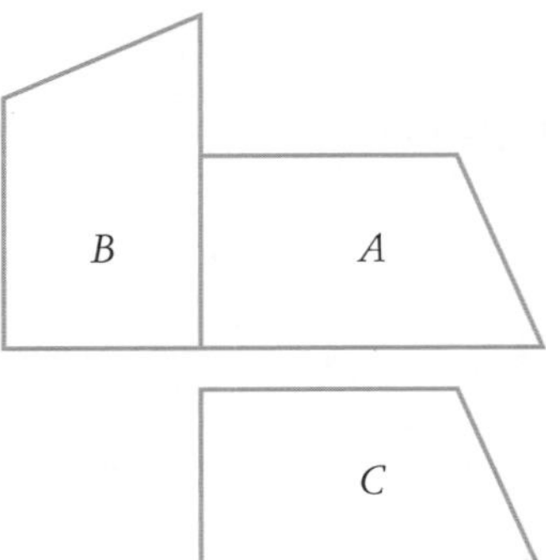

a to give Shape B? ________

b to give Shape C? ________

5 Explain how this shape does not have rotational symmetry.

6 What anti-clockwise rotation gives the same result as:

a a 90° clockwise rotation? ________

b a 180° clockwise rotation? ________

C S F

HW HOMEWORK

9780170454452

Name:
Due date:
Parent's signature:

Part A	/ 8 marks
Part B	/ 8 marks
Part C	/ 8 marks
Part D	/ 8 marks
Total	/ 32 marks

PART A: MENTAL MATHS

Calculators not allowed.

1 Evaluate $-4 \times 3 + (-2)$.

2 What angle is supplementary to 55°?

3 Convert $\frac{1}{5}$ to a percentage.

4 Evaluate $\frac{2}{3} - \frac{4}{7}$.

5 Simplify $4 \times a \times 3 \times y$.

6 Find the lowest common multiple of 9 and 3.

7 Evaluate 4.85×6.

8 What is the time 1 hour 15 minutes before 12:30 p.m.?

C S F

PART B: REVIEW

1 Draw an equilateral triangle.

2 On your diagram above, draw all axes of symmetry.

3 What is the order of rotational symmetry of an equilateral triangle?

4 Name each type of triangle.

a

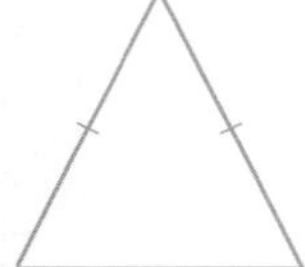

b

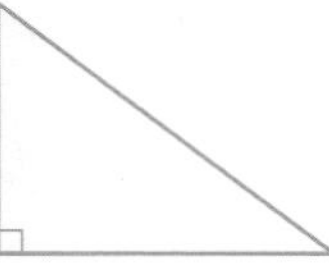

5 What type of triangle has:

a no equal sides?

b one axis of symmetry?

6 Draw an obtuse-angled triangle.

HW HOMEWORK

PART C: PRACTICE

› Classifying triangles
› Angle sum of a triangle
› Exterior angle of a triangle

1 Classify each triangle by sides and angles.

a

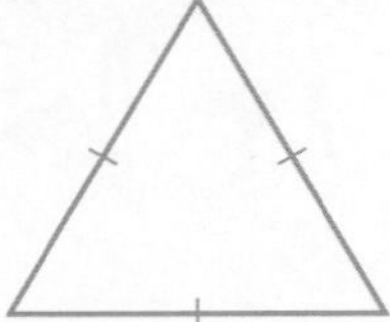

____

____

b

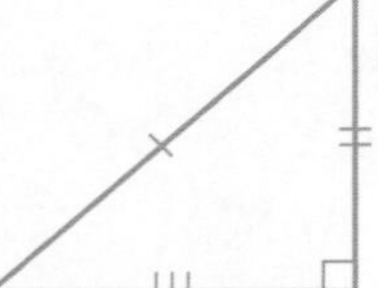

____

____

2 Draw an isosceles, right-angled triangle.

3 Find the value of each variable.

a

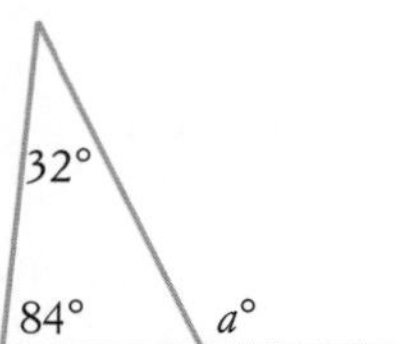

b

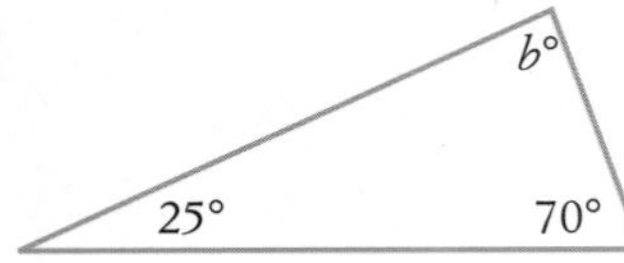

c

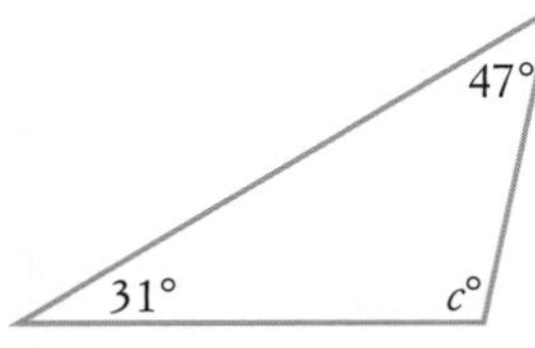

d

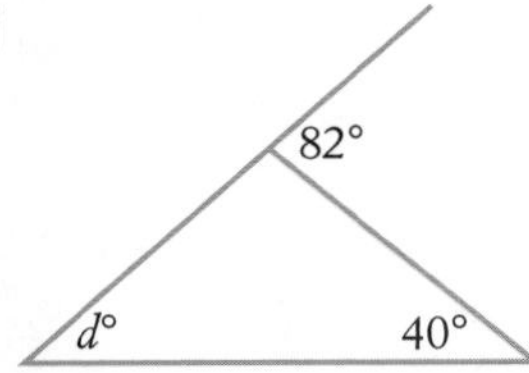

e

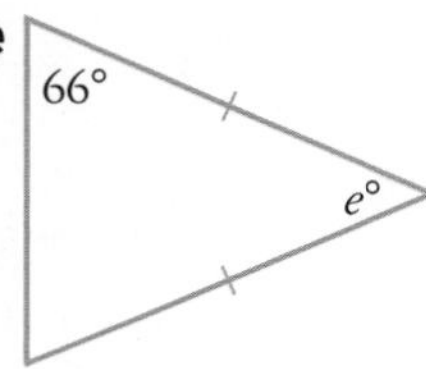

HW HOMEWORK

C S F

PART D: NUMERACY AND LITERACY

1 Name the 3 types of triangles if classifying by their angles.

____

____

2 What is the angle sum of any triangle?

____

3 For an equilateral triangle, find the size of:

a each angle ______________

b an exterior angle ______________

4 Complete: The exterior angle of a ______________ is equal to the ______________ of the 2 interior ______________ angles.

5 A triangle has a right angle. What is the sum of the other 2 angles? ______________

6 What is an acute-angled triangle?

____

____

7 What is the name of the angle created when a side of a triangle is extended?

____

HW HOMEWORK

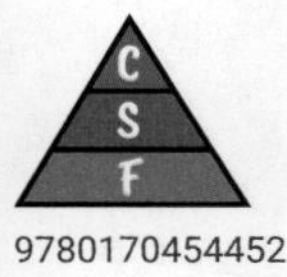

6 QUADRILATERALS

Name:

Due date:

Parent's signature:

Part A	/ 8 marks
Part B	/ 8 marks
Part C	/ 8 marks
Part D	/ 8 marks
Total	/ 32 marks

PART A: MENTAL MATHS

Calculators not allowed.

1 Evaluate 24×5.

2 Mark a pair of vertically opposite angles.

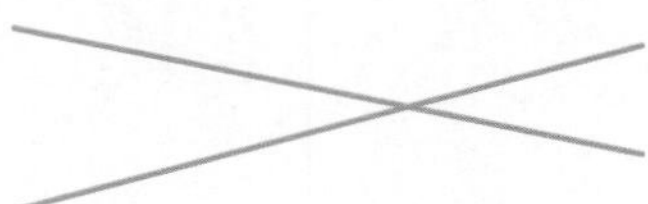

3 Test whether 708 is divisible by 9.

4 Convert 0.625 to a percentage.

5 Solve $6n + 18 = -12$.

6 Evaluate $\frac{4}{7} \times \frac{3}{8}$.

7 Complete: 8.4 kg = ______ g.

8 Evaluate $18 \div (-3) + 6$.

PART B: REVIEW

1 Draw a pair of parallel lines crossed by a transversal, then mark a pair of cointerior angles.

2 Find y.

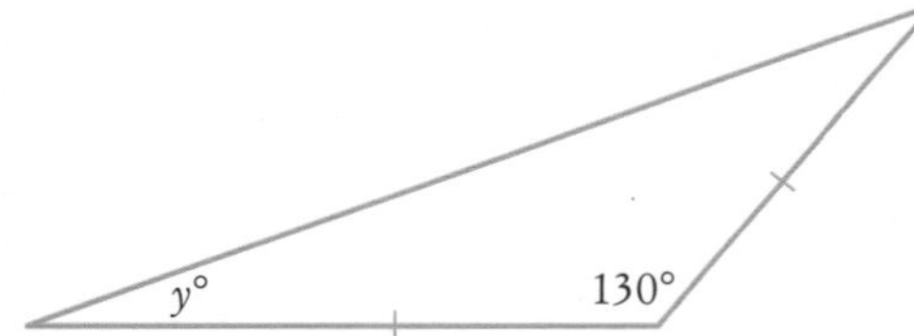

3 (2 marks) Classify the above triangle by sides and angles.

9780170454452

4 Name each shape.

a

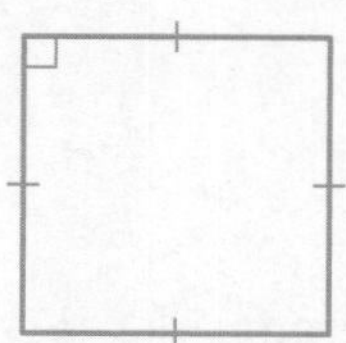

b

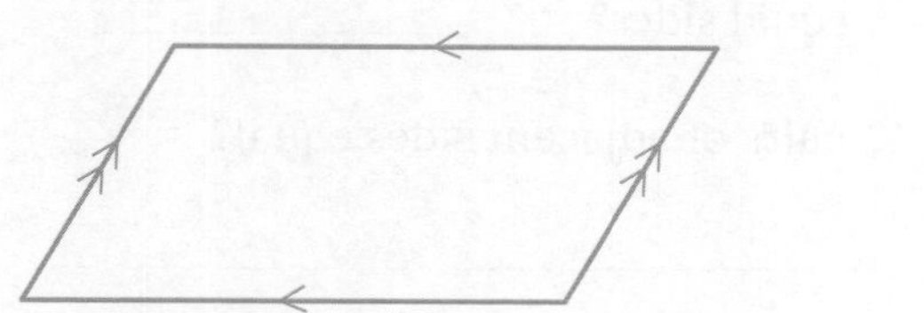

c

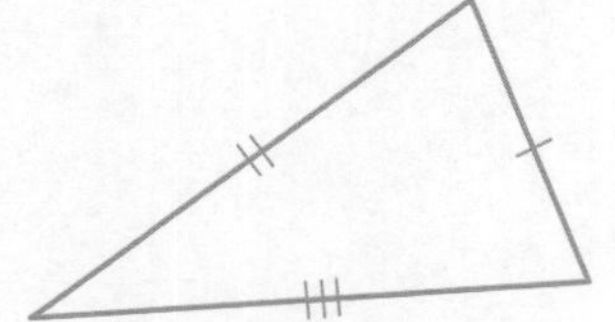

d

PART C: PRACTICE

› Classifying quadrilaterals
› Angle sum of a quadrilateral

1 Draw a rectangle, showing all features.

2 Find x and y.

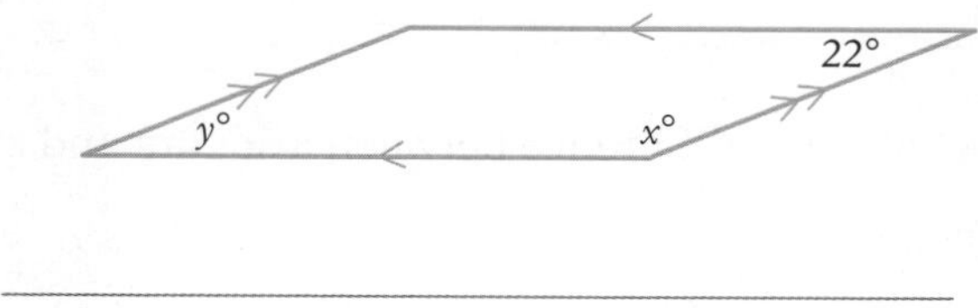

3 How many axes of symmetry has:

a a trapezium?

b a kite?

4 Draw a rhombus.

5 What order of rotational symmetry has a rhombus?

6 Find h.

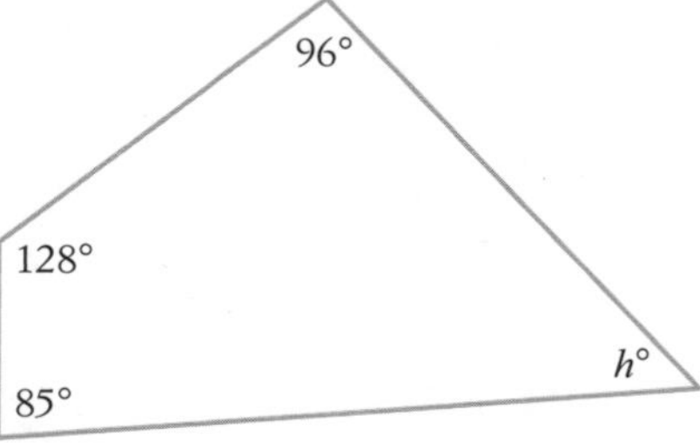

HW HOMEWORK

PART D: NUMERACY AND LITERACY

1 What is a quadrilateral?

____

____

2 What is the angle sum of any quadrilateral?

____

3 What is the difference between a square and a rectangle?

____

____

____

____

4 (2 marks) Circle all quadrilaterals that have both pairs of opposite sides equal:

square rectangle

trapezium kite

rhombus parallelogram

5 Which is the most general quadrilateral to have:

a 4 equal angles? ____________

b 4 equal sides? ____________

c 2 pairs of adjacent sides equal?

____

9780170454452

Name:

Due date:

Parent's signature:

Part A	/ 8 marks
Part B	/ 8 marks
Part C	/ 8 marks
Part D	/ 8 marks
Total	/ 32 marks

GEOMETRY REVISION

PART A: MENTAL MATHS

Calculators not allowed.

1 Evaluate 10^4.

2 Mark a pair of alternate angles.

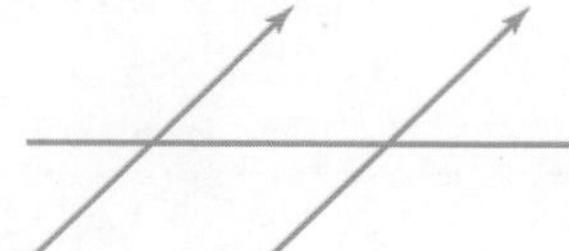

3 Find 30% of \$150.

4 **a** Write an algebraic expression for 'double a number n, less 8'.

b If double n less 8 equals 16, what is the value of n?

5 Evaluate $-16 \div (-8)$.

6 Estimate the value of 420×18.

7 How do you test whether 684 is divisible by 6?

PART B: REVIEW

1 (2 marks) Rotate this trapezium 90° anticlockwise about P, then translate it 4 units right, 2 units up.

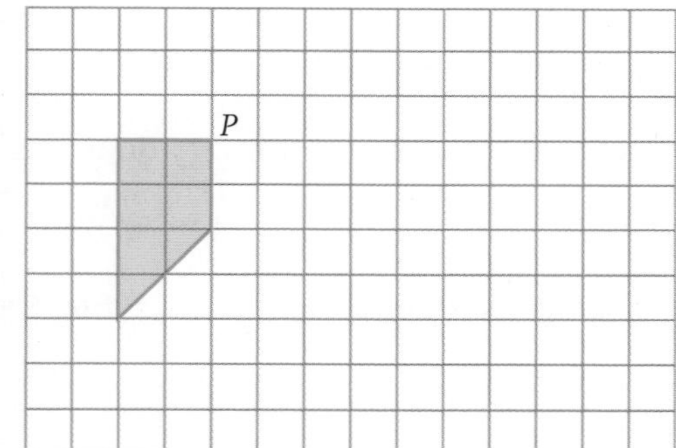

2 **a** What is a square?

b Draw a square and mark all axes of symmetry.

c State its order of rotational symmetry.

HW HOMEWORK

C S F

3 **a** (2 marks) Name the 2 transformations that have been performed on the triangle below.

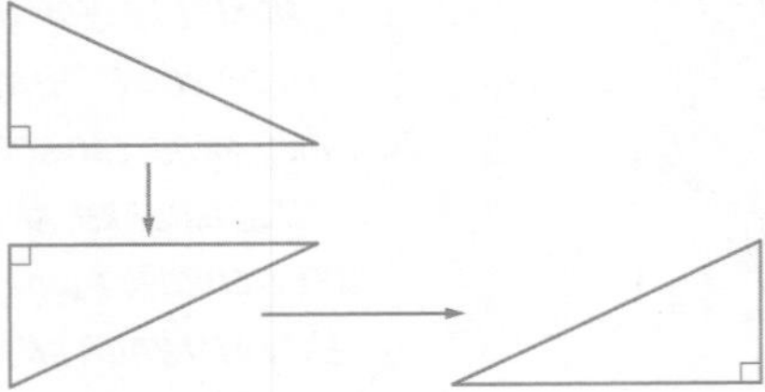

b Name one transformation that would give the same result as the 2 transformations.

____

PART C: PRACTICE

1 **a** Classify this triangle by sides and by angles.

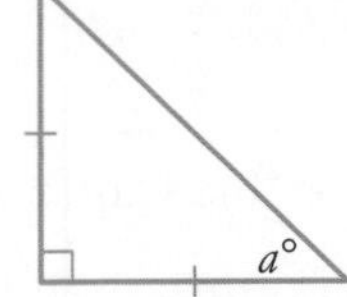

____

b Find the value of a. ______________________

c How many axes of symmetry does this triangle have? ______________________

d What is its order of rotational symmetry?

____

2 (4 marks) Find the value of each variable.

a

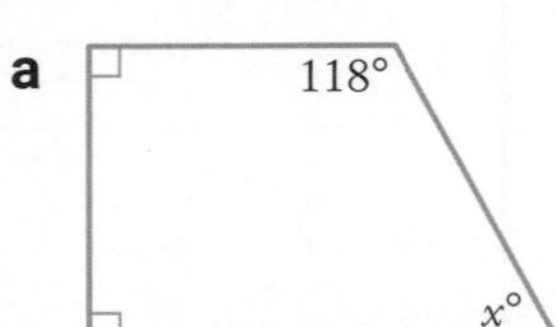

____

b

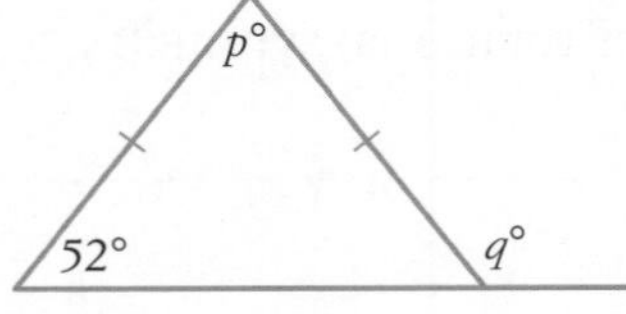

____

c

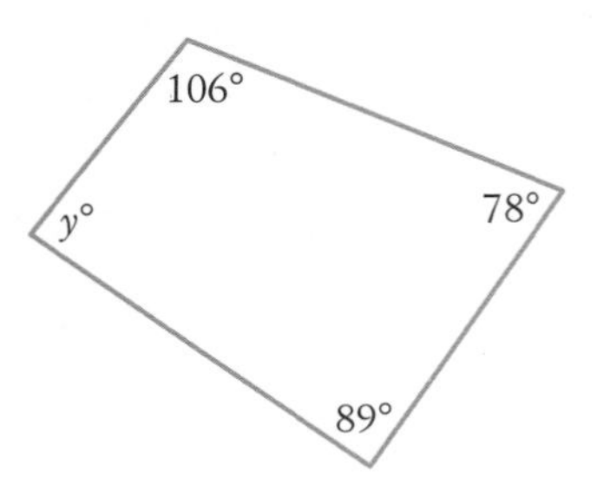

____

PART D: NUMERACY AND LITERACY

1 What is a rhombus?

____

____

2 (2 marks) State 2 properties of an equilateral triangle.

____

____

3 **a** Sketch a triangle showing 2 angle sizes whose sum is less than 90°.

b Find the size of the third angle.

____

c What type of triangle is it, if classifying by angles?

____

4 Which quadrilateral:

a has one axis of symmetry?

____

b has opposite sides equal and all angles 90°?

____

C S F

9780170454452

STARTUP ASSIGNMENT 7 

THIS ASSIGNMENT REVISES PREVIOUS WORK ON DECIMALS AND WHOLE NUMBERS TO PREPARE YOU FOR THE DECIMALS TOPIC.

WS WORKSHEET

PART A: BASIC SKILLS

/ 15 marks

1 Evaluate:

a $-3 + 5 - 8$ ____________

b $10 \times 8 \times 5$ ____________

c $27 - 12 \div 3 \times 2$ ____________

2 List the factors of 11. ____________

3 **a** Draw a trapezium.

b How many equal sides has a trapezium?

____

4 In this diagram why is $AB \parallel CD$?

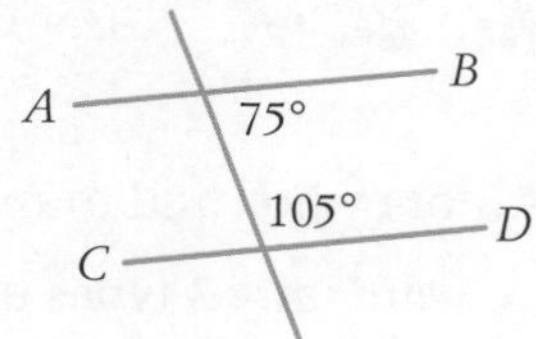

____

5 Write 4^3 in expanded notation. ____________

6 Convert $\frac{19}{4}$ to a mixed numeral. ____________

7 Find the lowest common multiple of 5 and 8.

____

8 **a** Draw a reflex angle.

b Label the reflex angle $\angle MPS$.

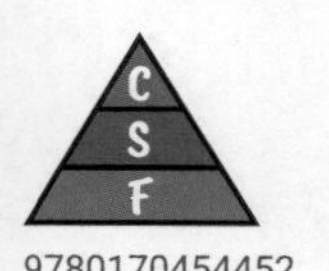

9 Simplify $3 \times r \times 7 \times d$. ____________

10 What is the size of each angle in an equilateral triangle? ____________

11 How do you test whether a number is divisible by 5?

____

PART B: NUMBER

/ 25 marks

12 Evaluate:

a 7×80 ____________

b 8×12 ____________

c $423 + 519 + 77$

____

____

d $290 - 147$

____

____

13 Complete this number line.

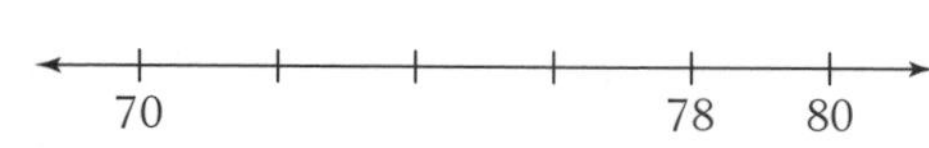

14 Write these numbers in descending order:

170, 117, 171, 107, 177, 111

____

15 What fraction of this shape is shaded?

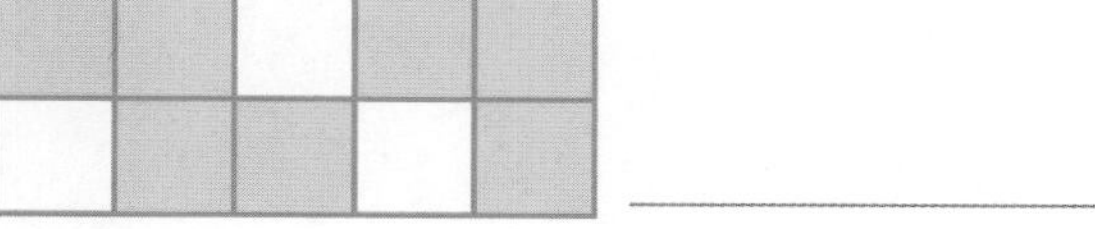

____

WS WORKSHEET

16 Write these numbers in ascending order:

354, 355, 534, 344, 543, 345

____

17 Evaluate:

a 19×100 ______________________

b $840 \div 10$ ______________________

c 284×3

____

____

d 135×26

____

____

18 In the numeral 4129:

a which digit is in the tens place? __________

b what is the value of the digit 1? __________

19 Round 75 238 to:

a the nearest hundred? __________

b the nearest thousand? __________

20 Simplify $\frac{45}{100}$. ______________________

21 Find:

a $1051 - 238$

____

____

b $2487 \div 3$ ______________________

c $1096 \div 8$

____

____

d $400 + 30 + 50$ ______________________

e 60×300 ______________________

22 Place $>$ or $<$ in the spaces to make each statement true.

a 37 ______ 74

b 8×2 ______ $10 - 6$

c $20 \div 2$ ______ 4^2

PART C: CHALLENGE

Bonus / 3 marks

Eddie emptied his money box and found that he had 50 coins. There were only 3 types of coins: $1, 20c and 10c. If the total value of the coins was $15.50, how many of each type of coin were there?

C S F

9780170454452

WHERE'S THE POINT? 7

WS WORKSHEET

This multiplication table shows each number in the top row multiplied by each number in the left column.

×	434	18	57
8	3472	144	456
23	9982	414	1311
169	73 346	3042	9633

1 Use the table to solve the following problems *without* using a calculator.

a 57 × 80 ______

b 169 × 570 ______

c 230 × 180 ______

d 4340 × 800 ______

e 1690 × 1800 ______

f 23 × 5.7 ______

g 169 × 43.4 ______

h 8 × 0.57 ______

i 2.3 × 1.8 ______

j 80 × 1.8 ______

k 4.34 × 169 ______

l 2.3 × 0.57 ______

m 2.3 × 570 ______

n 0.8 × 4.34 ______

o 34 720 ÷ 8 ______

p 3472 ÷ 80 ______

q 131 100 ÷ 57 ______

r 1311 ÷ 5.7 ______

s 733.46 ÷ 169 ______

t 9982 ÷ 230 ______

2 a Check your answers for question **1** using a calculator, and score a mark for each correct answer.

b How did you determine where the decimal point goes? Discuss the ways you worked out your answers.

3 Try to improve your score on the following set of questions using the multiplication table below.

×	16	315	24
49	784	15 435	1176
5	80	1575	120
131	2096	41 265	3144

a 49 × 2.4 ______

b 131 × 3.15 ______

c 5 × 0.16 ______

d 2.4 × 50 ______

e 31.5 × 500 ______

f 49 × 160 ______

g 50 × 3150 ______

h 0.49 × 2.4 ______

i 16 × 0.05 ______

j 13.1 × 31.5 ______

k 4.9 × 240 ______

l 160 × 1310 ______

m 160 × 13.1 ______

WS WORKSHEET

n 2.4 × 4.9 ______

o 0.5 ÷ 3.15 ______

p 41.265 ÷ 131 ______

q 8000 ÷ 16 ______

r 117.6 ÷ 49 ______

s 1543.5 ÷ 3.15 ______

t 31.44 ÷ 2.4 ______

4 a Check your answers for question **3** using a calculator.

b Did your score improve compared to question **1**?

______

c Which type of question was most difficult for you?

______

d Use the table in question **3** to help you write 5 more multiplication and/or division equations.

______

______

______

______

______

9780170454452

MULTIPLICATION ESTIMATION GAME 7

This is a game for 2 to 5 players or teams.

1. All players or teams have 2 minutes to estimate the answers to all 10 multiplication questions in the table.
2. After the time is up, 1 or 2 students use a calculator to find the exact answer to each question.
3. For each answer, score:
   - 5 points for each estimate within 2 of the correct answer
   - 4 points for each estimate within 4 of the correct answer
   - 3 points for each estimate within 6 of the correct answer
   - 2 points for each estimate within 8 of the correct answer
   - 1 point for each estimate within 10 of the correct answer
   - 0 points if the estimate is more than 10 from the correct answer.
4. The winner is the player or team with the most points.

	Multiplication	Estimation	Correct answer from calculator	Points
a	42×3.5			
b	5.1×2.4			
c	1.6×7.3			
d	3.8×6.2			
e	2.2×8.8			
f	7.5×3.2			
g	5.6×5.7			
h	2.6×5.1			
i	9.3×3.9			
j	4.8×3.3			
			Total	

INVESTIGATE FURTHER

1. Did you tend to overestimate or underestimate?
2. Now make up and try your own set of 10 decimal multiplications. See if you perform better with practice.

7 DECIMALS 1

Name:

Due date:

Parent's signature:

Part A	/ 8 marks
Part B	/ 8 marks
Part C	/ 8 marks
Part D	/ 8 marks
Total	/ 32 marks

PART A: MENTAL MATHS

Calculators are not allowed.

1 Evaluate 5^3.

2 Complete: 845 mL = ________ L

3 (2 marks) Find t and u.

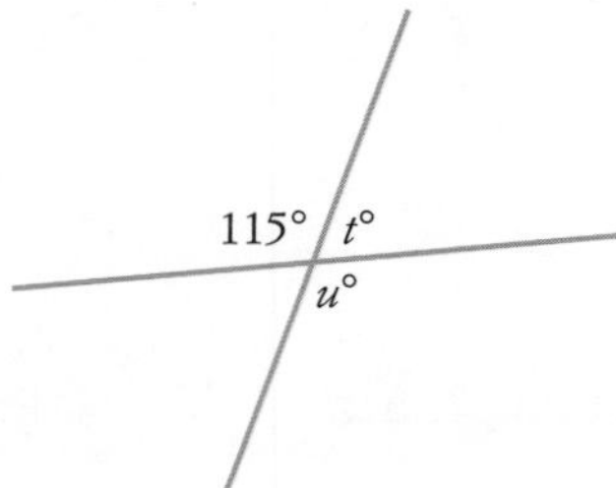

4 Evaluate 286 × 24.

5 Find the highest common factor of 30 and 75.

6 Complete: An equilateral triangle has

7 A square has an area of 64 cm^2.
What is its **perimeter**?

PART B: REVIEW

1 Evaluate 159 − 87.

2 List 66, 60, 640, 63, 6 in descending order.

3 Evaluate 31 × 100.

4 Evaluate 1800 ÷ 10.

5 Simplify $\frac{12}{100}$.

6 Round 3975 to the nearest:

a ten

b hundred

7 Simplify $\frac{245}{1000}$.

C S F

HW HOMEWORK

PART C: PRACTICE

› Ordering decimals
› Converting decimals to fractions
› Adding and subtracting decimals

1 Write $\frac{7}{1000}$ as a decimal.

2 Evaluate:

a 4.6 + 8.52

b 15.37 + 6.9

c 131.78 − 4.2

d 10.5 − 3.75

3 Write down a decimal between 4.16 and 4.2.

4 Complete:

a 2.54 + ________ = 10.7

b 13.01 − ________ = 9.8

PART D: NUMERACY AND LITERACY

1 **a** How many decimal places has 6.084?

b Which digit in 6.084 shows tenths?

2 The results of a long jump were Jack 8.06 m, James 7.48 m, Jason 7.87 m, Jay 8.1 m.

a Who jumped the furthest?

b Find the difference between Jack's and James' jumps.

c If Jules jumped 0.35 m further than Jason, what distance did Jules jump?

3 Convert 0.36 to a simple fraction. Show working.

4 (2 marks) Place these library books in ascending order of their call numbers: *The Snooker Book* 794.735; *Chess Made Easy* 794.109; *Mah Jong Explained* 795.34; *Ten-pin Bowling for Winners* 794.6.

HW HOMEWORK

7 DECIMALS 2

THIS ASSIGNMENT PRACTISES MULTIPLYING AND DIVIDING DECIMALS. MAKE SURE YOU LEARN AND REMEMBER THE SPECIAL STRATEGIES BEHIND THESE.

Name:

Due date:

Parent's signature:

Part A	/ 8 marks
Part B	/ 8 marks
Part C	/ 8 marks
Part D	/ 8 marks
Total	/ 32 marks

PART A: MENTAL MATHS

Calculators are not allowed.

1 Evaluate $25 \times 9 \times 4 \times 2$.

2 Write 6:35 p.m. in 24-hour time.

3 (2 marks) Classify this triangle by sides and angles.

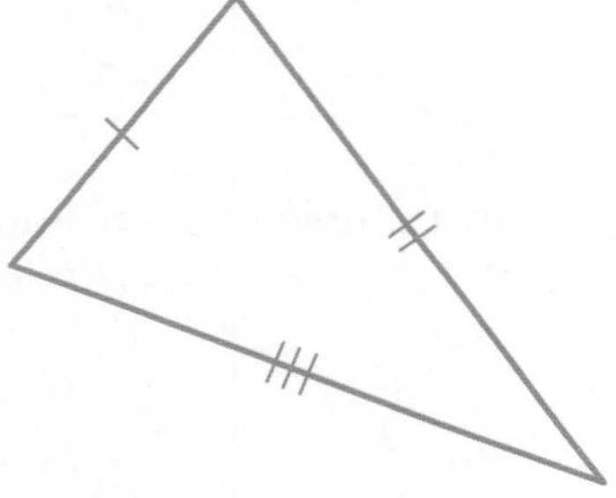

4 Find the square number between 60 and 70.

5 Simplify $5r - r$.

6 How many faces has a rectangular prism?

7 Find the lowest common multiple of 25 and 60.

PART B: REVIEW

1 Complete $\frac{32}{100} = \frac{}{1000}$.

2 Convert $\frac{32}{100}$ to a decimal.

3 Evaluate $8.32 + 0.9 + 12.54$.

4 Arrange 7.35, 7.04, 7.477, 7.8, 7.3 in ascending order.

5 Convert 0.65 to a simple fraction.

6 Evaluate

a 8400×10

b $8400 \div 100$

7 Evaluate $35.1 - 2.95$.

C S F

9780170454452

PART C: PRACTICE

› Multiplying decimals
› Dividing decimals

1 Evaluate

a $45.62 \times 10\,000$

______

b $86.4 \div 100$

______

c 135.8×12

______

d 0.4×0.007

______

e 2.9×0.06

______

f $68.45 \div 5$

______

g $8.642 \div 0.8$

______

h $284.4 \div 0.06$

______

PART D: NUMERACY AND LITERACY

1 Complete: To divide a decimal by 10, move the decimal point

______

2 Marcus swam 50 metres in the following times (in seconds): 38.1, 38.5, 40.7, 37.2, 36.0

a What is the difference between his longest and shortest times?

______

b Find the sum of Marcus' 5 times.

______

c Find the average of Marcus' 5 times.

______

3 If $257 \times 13 = 3341$, evaluate:

a 2.57×1.3 ______

b 257×0.13 ______

4 (2 marks) Complete: To divide a decimal by a decimal, make the second decimal a ______ number by moving the decimal point the required number of places to the ______.

For example:

$4.168 \div 0.08 =$ ______ $\div$ ______

$=$ ______

7 DECIMALS 3

Name:

Due date:

Parent's signature:

Part A	/ 8 marks
Part B	/ 8 marks
Part C	/ 8 marks
Part D	/ 8 marks
Total	/ 32 marks

PART A: MENTAL MATHS

Calculators are not allowed.

1 Evaluate $\sqrt{16}$.

2 Complete: $8 \times 6 =$ ________ $\times 12$

3 List all the prime numbers between 12 and 18.

4 Name this solid.

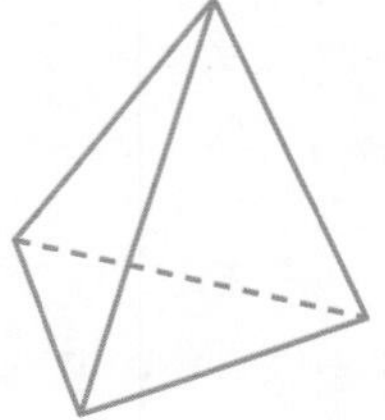

5 If $x = 8$, evaluate $9 + 4x$.

6 Complete: 6280 cm = ________ m.

7 Evaluate $2.16 \times 10\,000$.

8 If the perimeter of a square is 12 m, what is the length of each side?

PART B: REVIEW

1 Arrange 10.061, 10.6, 10.07, 10.63, 10.3 in ascending order.

2 Convert 0.38 to a simple fraction.

3 Evaluate 16.7×18.

4 Complete: $4.095 \div 0.18 =$ ________ $\div 18$

$=$ ________

5 A bottle of soft drink costs \$3.25. How many full bottles can be bought for \$50?

6 Evaluate $308.9 \div 100$.

7 Given that $22^2 = 484$, evaluate:

a 2.2^2 ________

b $(0.22)^2$ ________

C S F

9780170454452

HW HOMEWORK

PART C: PRACTICE

› Terminating and recurring decimals
› Rounding decimals
› Decimal problems

1 Convert each fraction to a decimal.

a $\frac{3}{8}$ ______________________

b $\frac{5}{6}$ ______________________

2 Evaluate $25.76 \div 0.7$.

____

3 Round 41.6375 to:

a the nearest tenth ______________________

b 3 decimal places ______________________

4 The cost of 5 kg of honey is $19.75. Find the cost of 3 kg of honey.

____

____

____

5 Emma jumped 2.38 m in the long jump, which was 0.4 of Jenny's winning jump. How long was Jenny's jump?

____

____

____

6 Write a decimal that could be **rounded up** to 4.09.

____

PART D: NUMERACY AND LITERACY

1 **a** What type of decimal is $0.\dot{2}\dot{7}$?

____

b Write $0.\dot{2}\dot{7}$ the long way.

____

2 (3 marks) Complete: To round 195.748 to one decimal place, cut the decimal after the digit __________. The next digit is __________, so round __________ because it is __________ than 5. This gives the answer ______________________.

3 A builder bought 1000 bricks for $348.80 and 300 tiles at $1.85 per tile to build a patio.

a How much did each brick cost, to the nearest cent?

____

b How much did the tiles cost?

____

____

c What was the total cost of the bricks and tiles?

____

____

9780170454452

7 DECIMALS REVISION

HOW ARE YOU GOING WITH MASTERING DECIMALS? HAVE YOU BUILT UP YOUR STRENGTH AND CONFIDENCE?

Name:

Due date:

Parent's signature:

Part A	/ 8 marks
Part B	/ 8 marks
Part C	/ 8 marks
Part D	/ 8 marks
Total	/ 32 marks

PART A: MENTAL MATHS

Calculators are not allowed.

1 How many hours in 75% of one day?

2 If $d = 5$, evaluate $d^2 + d$.

3 List all the composite numbers below 10.

4 Find y.

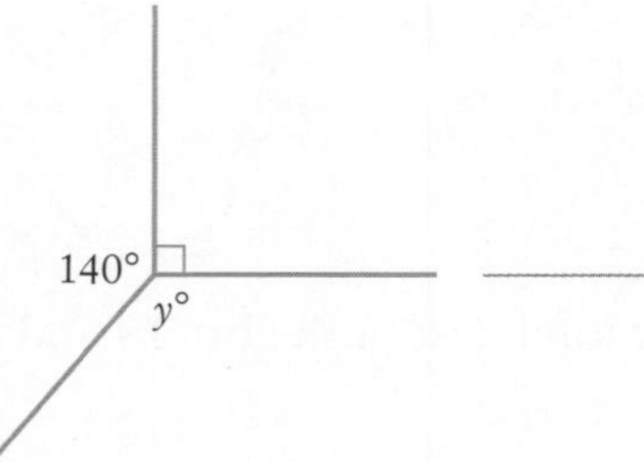

5 What is the probability that a baby is born on a day of the week beginning with S?

6 Complete: 58.2 L = ______ mL.

7 Solve $3n + 20 = 62$.

8 Find the perimeter of a square that has an area of 100 m^2.

PART B: REVIEW

1 Arrange 3.6, 3.46, 3.69, 3.7, 3.645 in ascending order.

2 Round 14.028 to the nearest hundredth.

3 Evaluate:

a 34.61×1000

b $6.258 \div 100$

c $276.4 + 53.98$

d $567.3 - 34.82$

e 23.6×0.03

f $452.84 \div 0.8$

C S F

HW HOMEWORK

9780170454452

PART C: PRACTICE

› Decimals revision

1 Convert each fraction to a decimal.

a $\frac{5}{16}$

b $\frac{4}{7}$

2 Evaluate $8.9 - 2.4 \div 2$.

3 Evaluate $(0.8)^3$.

4 Write a decimal that can be rounded to 25.1.

5 Randall went shopping and bought:

- 2 cartons of milk @ $2.20 each
- 3 kg bananas @ $2.19/kg
- 5 L of ice cream @ $1.85/L
- 6 apples @ 65c each

a Find the cost of the ice cream.

b Find the total cost of all items, rounded to the nearest 5 cents.

c Calculate Randall's change from $50.

PART D: NUMERACY AND LITERACY

1 (2 marks) What is the difference between a terminating and recurring decimal?

2 Write 0.316 316 316... using recurring decimal notation.

3 Find the product of 4.6 and 0.7.

4 Complete: Rounding to the nearest thousandth means rounding to ________ decimal places.

5 Julia saves $224.50 per week for 3 weeks but spends $32.80 per week for 4 weeks. How much money does she have left?

6 Find the difference between 56.4 and 8.32.

7 If petrol costs $1.51 per litre, how many litres (correct to one decimal place) can be bought for $60?

HW HOMEWORK

7 DECIMALS CROSSWORD

PS PUZZLE SHEET

Unscramble the words and place them in the crossword.

1 2 3 4 5 6 7 8 9 10 11 12 13

Across

5 NDCNEIDEGS

7 LPACE

8 TSAHTHDNOU

10 DORUN

11 UIRRNECGR

12 HDHRUENTD

13 ESATMETI

Down

1 TRNFOACI

2 IPOTN

3 DLAMECI

4 TEIMRNTINGA

6 NHETT

9 NANGCDSIE

9780170454452

STARTUP ASSIGNMENT 8

THIS ASSIGNMENT REVISES YOUR MEASUREMENT SKILLS TO PREPARE YOU FOR AREA AND VOLUME.

PART A: BASIC SKILLS / 15 marks

1 Simplify $3 \times m \times p \times p$. ______

2 Draw a 145° angle.

3 List the first 5 multiples of 7.

______

4 Solve $2x + 5 = 17$. ______

5 How many days in October? ______

6 **a** Draw a factor tree for 72. ______

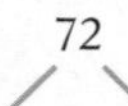

b Write 72 as a product of its prime factors.

______

7 Complete: For a number to be divisible by 6, it must be divisible by both ______ and ______.

8 What is the value of Q on this scale?

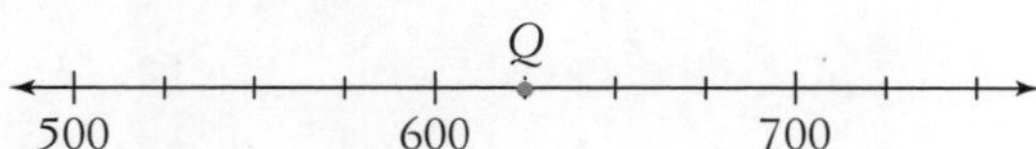

9 Complete with a > or <:

8.41 ______ 8.7

10 Find x.

______

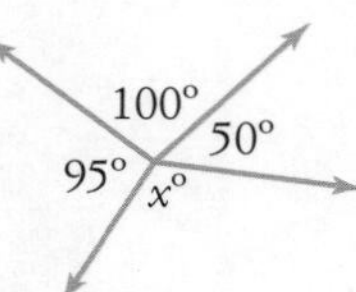

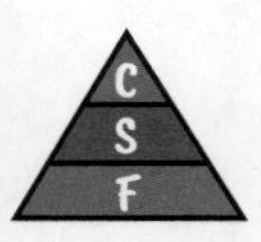

11 Evaluate:

a $-5 + (-7)$ ______

b 6^3 ______

c $\frac{2}{5} - \frac{1}{8}$ ______

12 If $a = -2$, then evaluate $3a - 7$. ______

PART B: MEASUREMENT / 25 marks

13 Complete:

a 1 m = ______ mm

b 1 km = ______ m

c 1 cm = ______ mm

d 1 L = ______ mL

14 What is the size of one unit on this ruler?

0 10 20 30 40
mm

15 For this rectangle, find:

a its perimeter ______

b its area ______

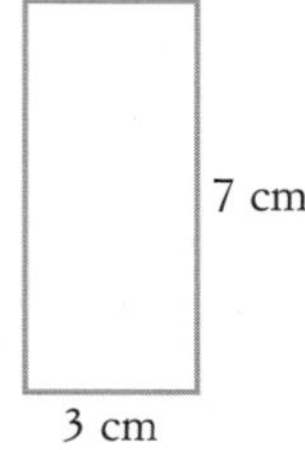

16 Evaluate:

a 38.52×10 ______

b 0.195×1000 ______

c $2074 \div 100$ ______

d $435.6 \div 100$ ______

17 A square has an area of 25 cm². What is its:

a side length? ______

b perimeter? ______

WS WORKSHEET

18 Name this solid.

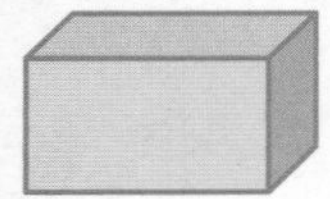

19 Find the area of this rectangle.

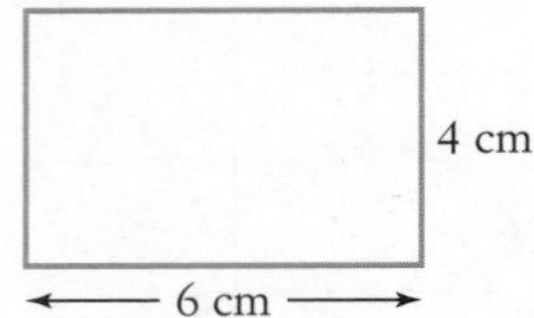

20 Evaluate:

a 0.5×2

b $\frac{1}{2} \times 7 \times 2$

c $470\,000 \div 10\,000$

21 **a** What type of quadrilateral is *ABCD*?

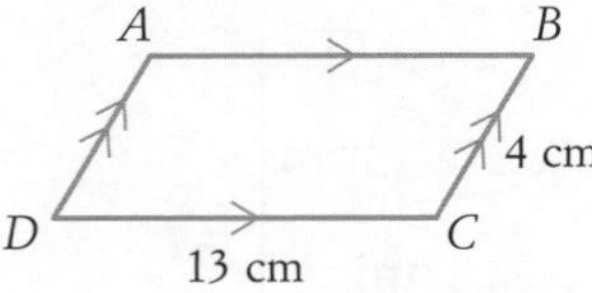

b Write the length of *AD*.

c Explain why $\angle A + \angle D = 180°$.

22 Evaluate:

a 17×100

b $4000 \div 10$

c $240 - 30$

d $150 - 45$

PART C: CHALLENGE

Bonus / 3 marks

Sammi, Charlotte, Tim and Jordan sat with Mum (M) and Dad (D) around the dinner table.

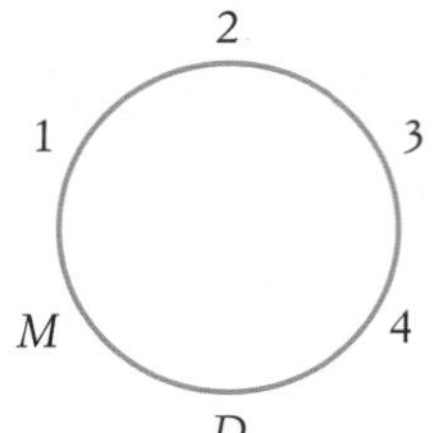

- Jordan sat with one of his parents.
- Tim sat between Sammi and Charlotte and opposite Dad.
- Sammi sat between Tim and Jordan.

Where did each person sit?

C S F

9780170454452

FIND THE AREA BY COUNTING SQUARES AND HALF-SQUARES.

WS WORKSHEET

Find the area of each figure if each square on the grid is 1 cm^2.

1 2 3 4 5 6 7 8 9 10 11 12 13 14 15 16 17 18

WS WORKSHEET

8 COMPOSITE AREAS

THESE COMPOSITE SHAPES ARE MADE UP OF 2 OR MORE BASIC SHAPES. THE PROBLEMS ARE NOT EASY, AND THE RED QUESTIONS NEXT PAGE ARE EXTRA CHALLENGING.

Find the shaded area of each figure.

1

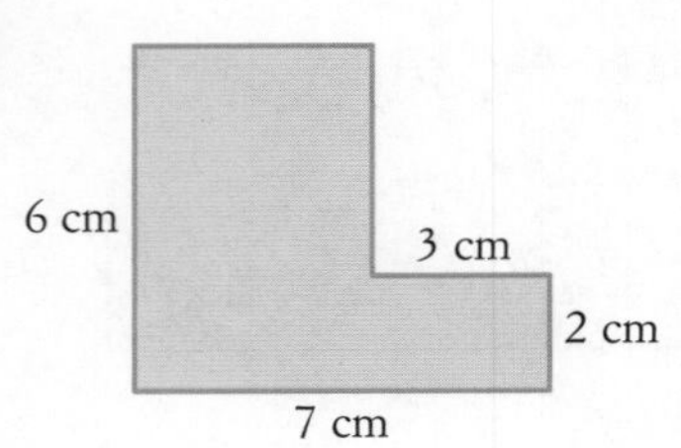

2

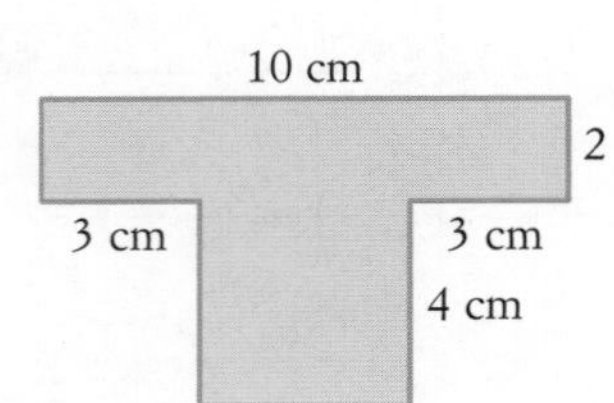

3

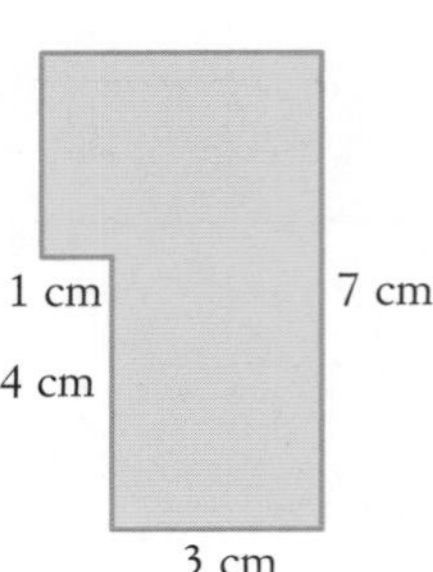

4

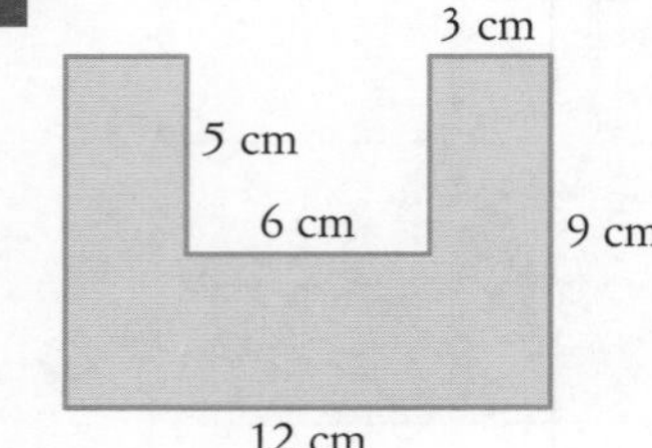

5

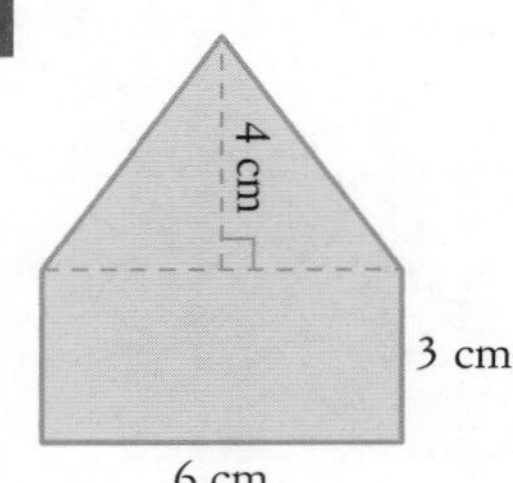

6

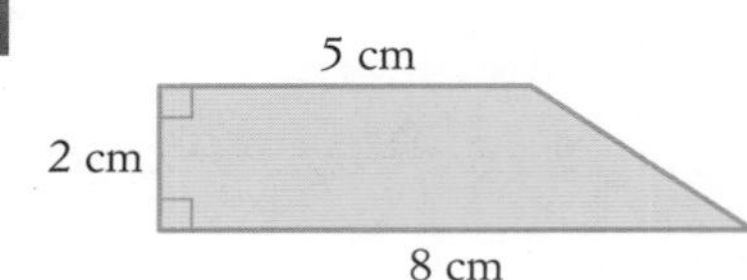

7

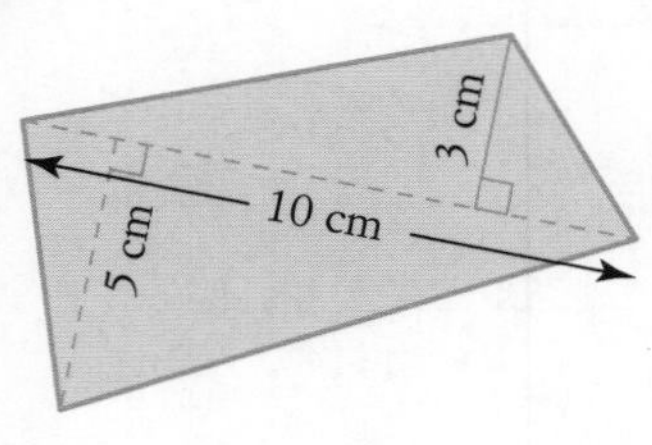

8

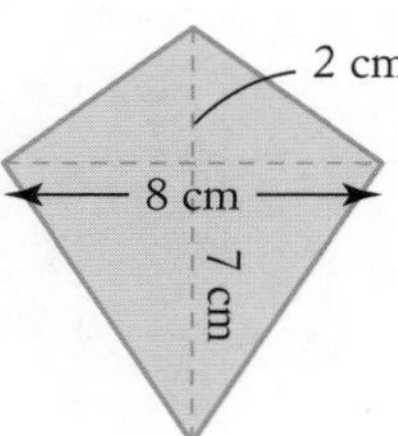

9

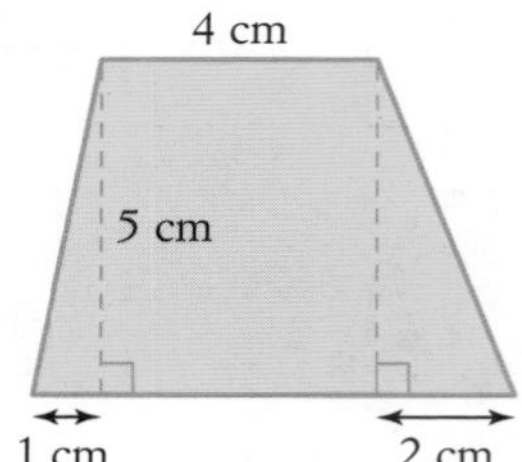

C
S
F

9780170454452

10

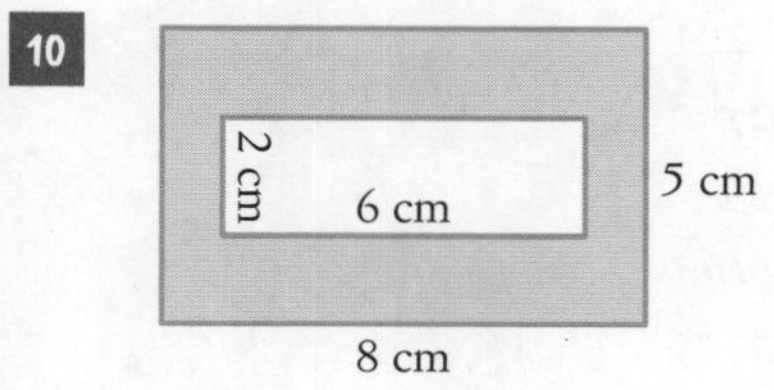

11

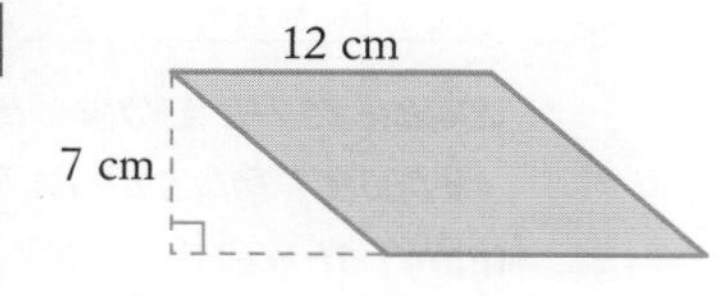

12

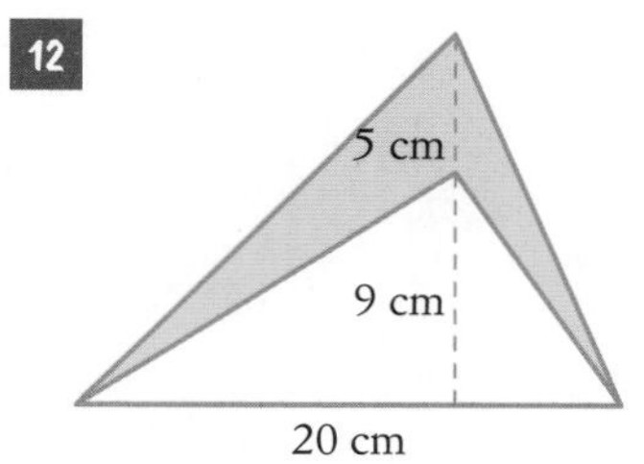

13

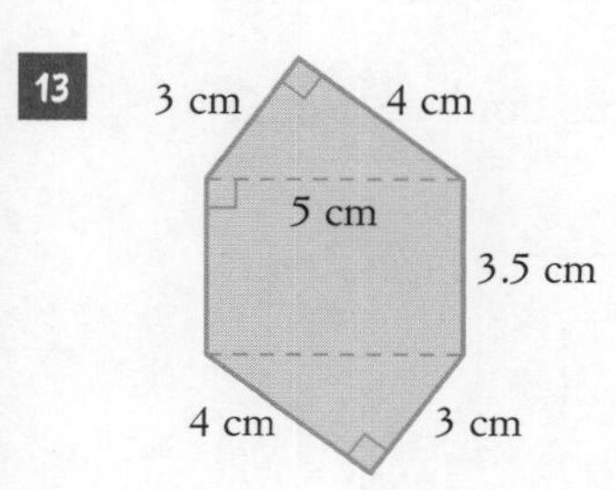

14

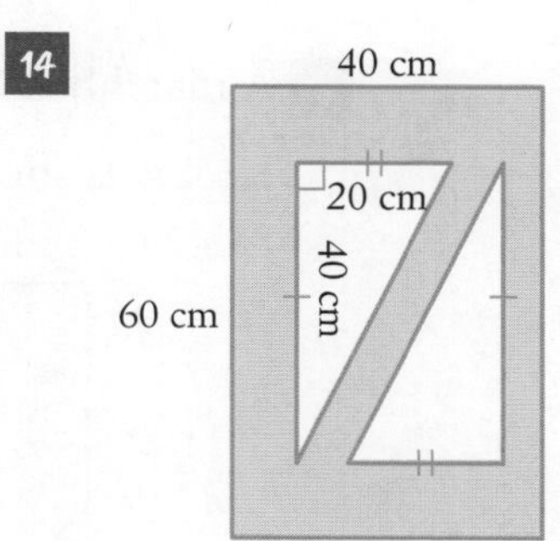

15

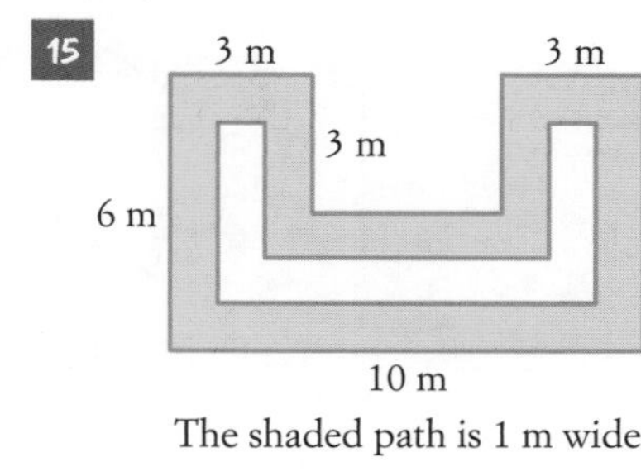

The shaded path is 1 m wide

16

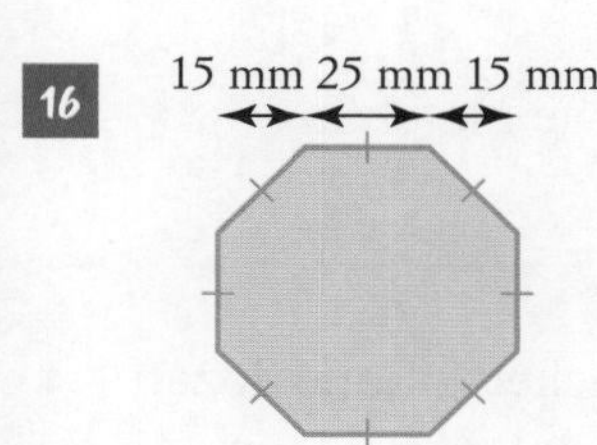

17

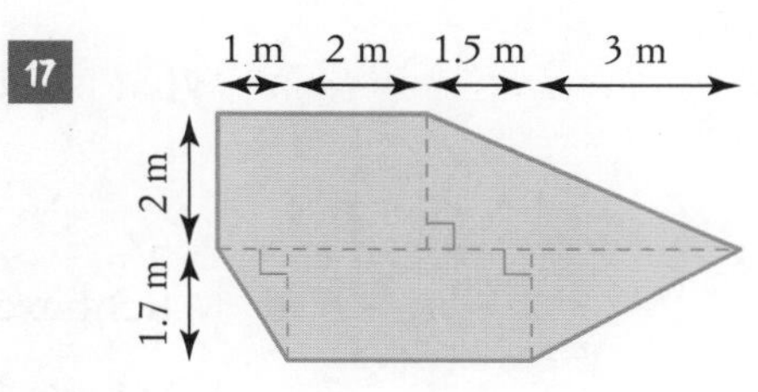

18

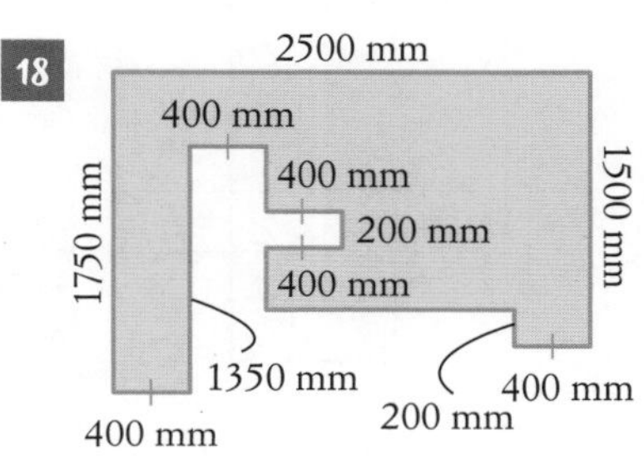

WS WORKSHEET

C
S
F

8 VOLUME AND CAPACITY

WS WORKSHEET

1 A 10 cm cube has a 5 cm square hole cut through the centre of it as shown. What is the volume of the remaining piece?

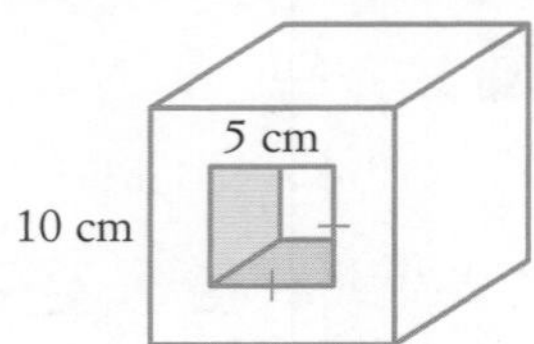

____

2 This is the net of a rectangular prism. What is the volume of the prism?

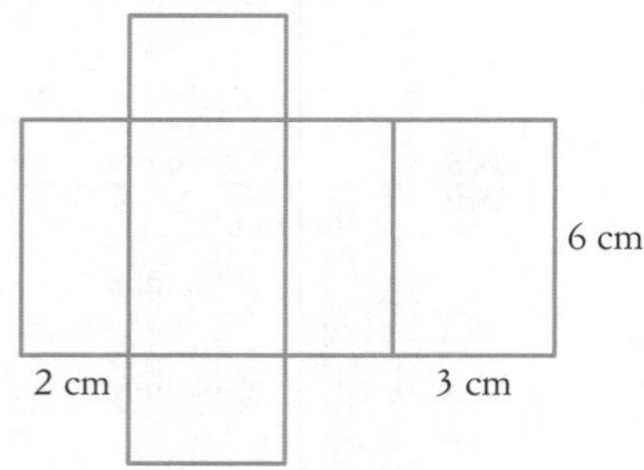

3 A rectangular piece of metal has 4 squares cut from the corners as shown. It is then folded to form an open rectangular prism with dimensions as shown.

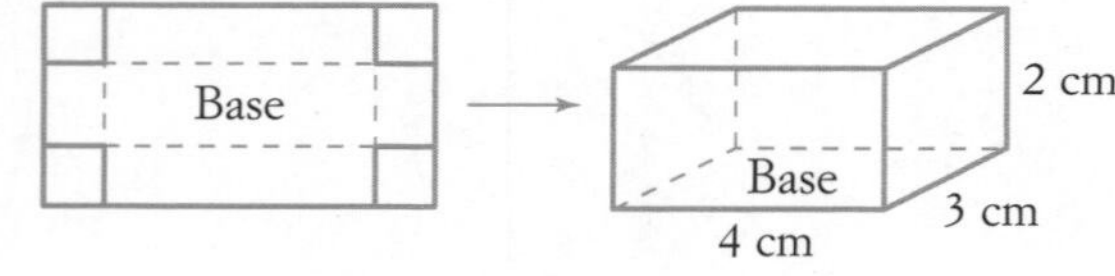

a What was the area of the original sheet of metal? ______________________

b What is the volume of the prism?

____

4 A swimming pool holds 16 000 L of water. The area of its base is 20 m^2. What is the depth of water in the pool?

____

5 A cube has a volume of 125 cm^3. What is the length of one side?

____

6 A matchbox measures 5.5 cm by 3.5 cm by 1.5 cm as shown.

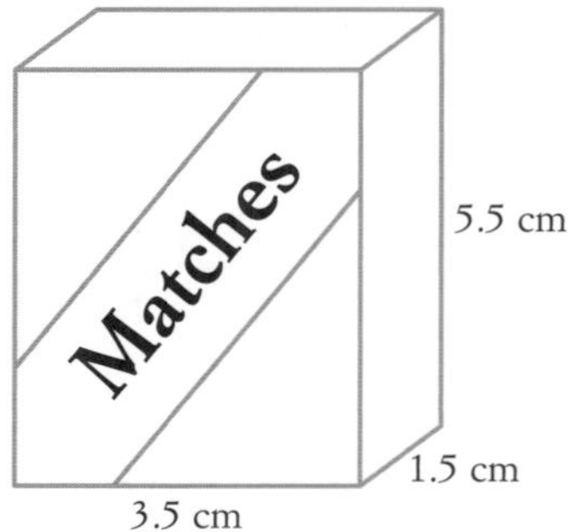

a What is the volume of the matchbox?

____

b Matchboxes come in packets of one dozen. What would the volume of one packet be?

____

c Find as many different ways as possible of making up the packet with 12 matchboxes.

____

____

d Which of these possibilities would need the least paper to package it?

____

7 An empty diving pool is in the shape of a rectangular prism with a base measuring 6 m by 4 m. Water flows into the pool at the rate of 800 L per hour. How deep is it after 5 hours?

____

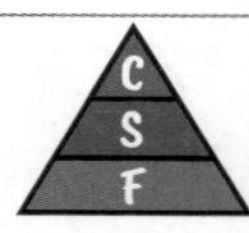

9780170454452

8 A steel cube with a side length of 3 cm has a mass of 210.6 grams. What is the mass of 1 cubic centimetre of this steel?

9 A water tank with the dimensions shown has an overflow pipe fitted 15 cm from the top. What is the greatest volume of water that the tank can hold?

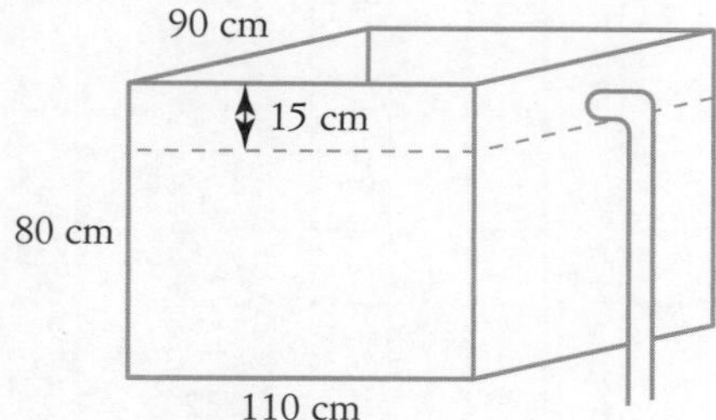

10 The edge of a cube box is 1 m. How many cubes of edge 2 cm will fit into the box?

11 A large cube is made up of 343 small cubes. How many of the small cubes are there along one of the edges of the large cube?

CHALLENGES

1 The area of each face of this rectangular prism is given. What is its volume?

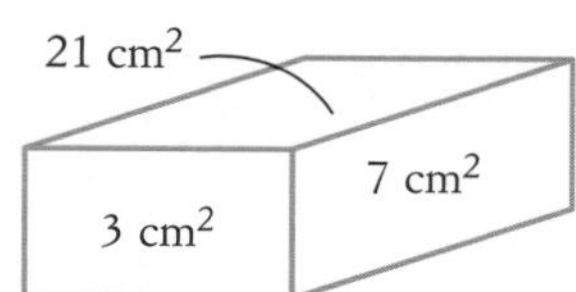

2 If the sides of a rectangular prism are increased by 2 cm, 3 cm and 4 cm, it becomes a cube and its volume is increased by 827 cm^3. Find the dimensions of the original prism.

WS WORKSHEET

8 AREA AND VOLUME CROSSWORD

THIS PUZZLE SHOWS THAT WORDS AND NUMBERS ARE BOTH IMPORTANT WHEN IT COMES TO SUCCESS IN MATHS.

1 2 3 4 5 6 7 8 9 10 11 12 13 14 15 16 17 18 19 20 21 22 23 24

Across

2 Starting with B, another word for 'width'

3 A rectangle whose length and width are the same

6 The amount of surface enclosed by a shape

8 A prefix represented by the Greek letter μ

14 Not length or width

16 Quadrilateral with opposite sides equal

19 Unit of capacity equal to 1 cubic metre

21 Metric prefix with abbreviation 'c'

22 Metric prefix meaning 'one thousandth'

23 How long something is

24 Solid shape with identical 'slices'

Down

1 At right angles

4 Its area is length × width.

5 Its area is $\frac{1}{2}$ × base × height.

7 Type of shape made up of 2 or more shapes

8 Prefix meaning 'one million'

9 The distance around the outside of a shape

10 The amount of material that a container can hold

11 Our system of measurement

12 The amount of space occupied by a solid

13 A 'slice' of a solid shape is called a ____-section.

15 1000 kg

17 1000 mm

18 A unit of capacity

19 Prefix meaning 'one thousand'

20 The area of a 100 m by 100 m square

21 Volume can be measured in ____ metres.

9780170454452

Name:

Due date:

Parent's signature:

Part A	/ 8 marks
Part B	/ 8 marks
Part C	/ 8 marks
Part D	/ 8 marks
Total	/ 32 marks

MEASUREMENT 8

PART A: MENTAL MATHS

Calculators are not allowed.

1 Evaluate $-3 \times 8 \div 2$.

____

2 What transformation involves a 'mirror-image'?

____

3 Convert 80% to a simple fraction.

____

4 Find w.

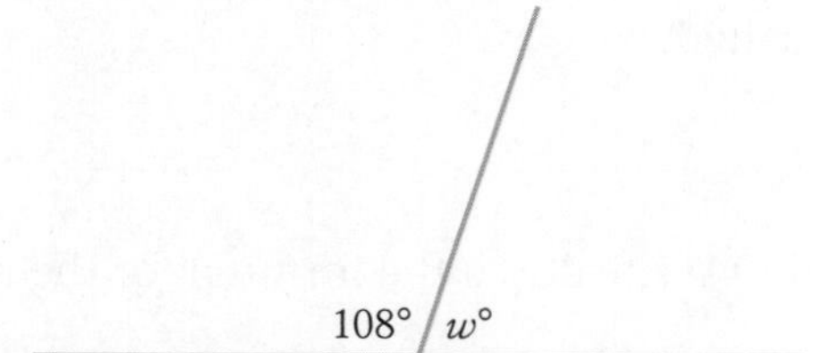

____

5 Evaluate 15×40.

____

6 Simplify $3 \times a \times 9 \times a \times b$.

____

7 Find the sum of the first 4 square numbers.

____

8 Evaluate $\frac{1}{4} - \frac{1}{5}$.

____

PART B: REVIEW

1 Complete:

a 5.1 t = ________ kg

b 0.78 m = ________ mm

c 96 h = ________ days

d 547 L = ________ kL

2 An equilateral triangle has sides of length 4 cm. What is its perimeter?

____

3 (2 marks) Find x and y.

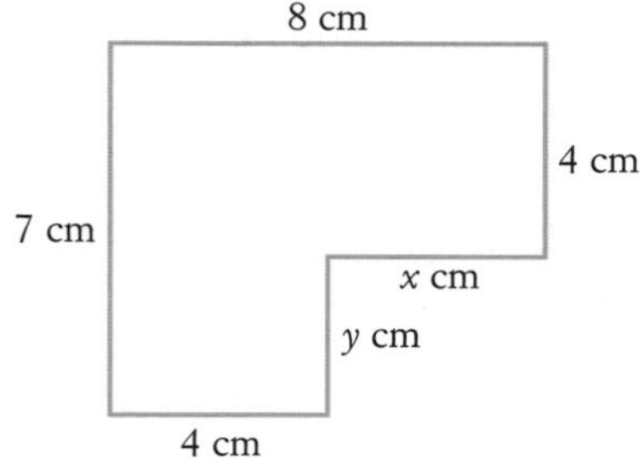

____

4 Which metric unit of area is approximately the size of a fingernail?

____

C S F

HW HOMEWORK

HW HOMEWORK

PART C: PRACTICE

› Metric units for area
› Perimeter
› Area of a rectangle

1 For this rectangle, find:

3.4 cm

7.9 cm

a its perimeter ____________________

b its area ____________________

2 Complete:

a $1 \text{ m}^2 =$ __________ mm^2

b $1 \text{ cm}^2 =$ __________ mm^2

3 Find the perimeter of this shape.

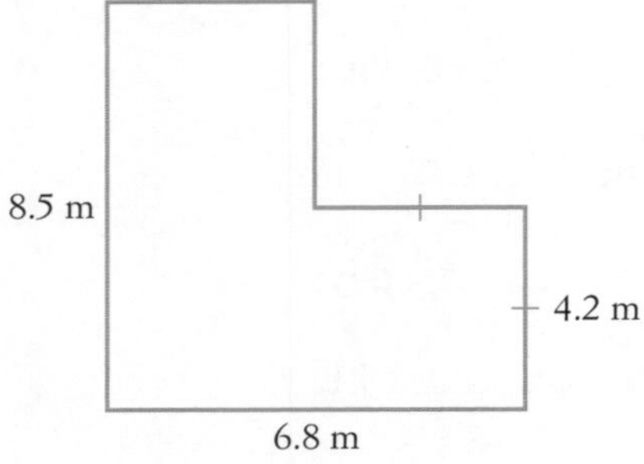

____

4 Cooper's rectangular bedroom floor measures 3.5 m by 4 m. He buys carpet squares of length 50 cm.

a What is the area of the bedroom floor?

____

b What is the area of one carpet square in square metres?

____

c How many carpet squares are needed to cover the bedroom floor?

____

PART D: NUMERACY AND LITERACY

1 Describe a square metre.

____

____

2 What is the formula '$P = 2l + 2w$' used for?

____

____

3 What does the metric prefix 'milli-' mean?

____

4 Which metric prefix means 'million'?

____

5 How do you say the metric unit of volume 'm^3' in words?

____

6 What is the sum of the lengths of the sides of a shape called?

____

7 (2 marks) Write down the formula for the area of a rectangle, describing what each variable stands for.

____

____

____

____

C S F

9780170454452

Name:

Due date:

Parent's signature:

Part A	/ 8 marks
Part B	/ 8 marks
Part C	/ 8 marks
Part D	/ 8 marks
Total	/ 32 marks

AREA 8

PART A: MENTAL MATHS

Calculators are not allowed.

1 Round 3516 to the nearest hundred. ________

2 Complete: $5 \times 12 =$ ________ $\div 10$.

3 Find a and b.

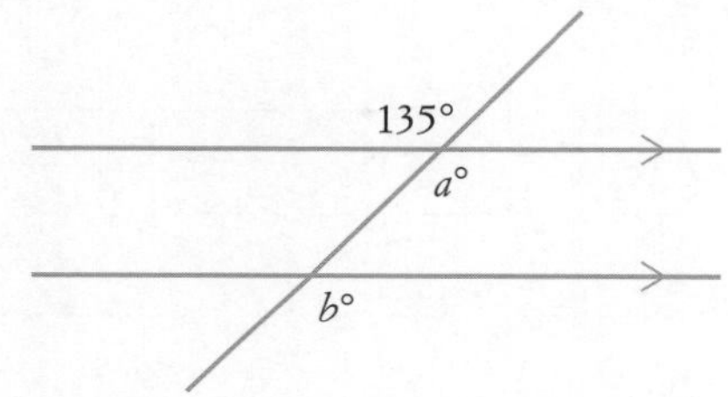

____

4 Evaluate $3 + (-8) - (-4)$.

____

5 Solve $2c - 21 = 7$.

____

____

____

6 Complete: 5.6 g = ________ mg.

7 Draw a triangular prism.

C
S
F

PART B: REVIEW

1 Find the perimeter of each shape.

a

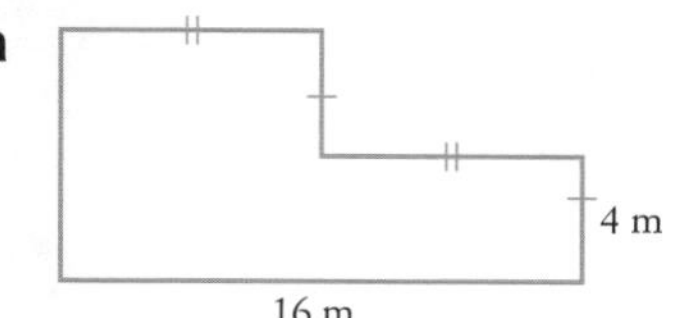

____

b

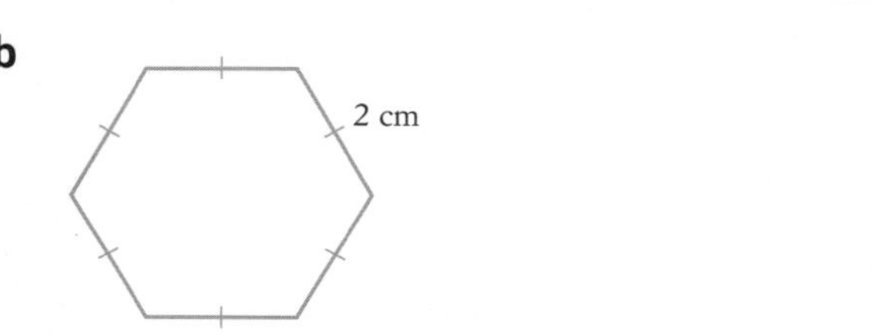

____

2 Complete: 640 mL = ________ L.

3 A square has a perimeter of 18 m. Find:

a the length of one side

____

b its area

____

4 Find the area of each rectangle.

a

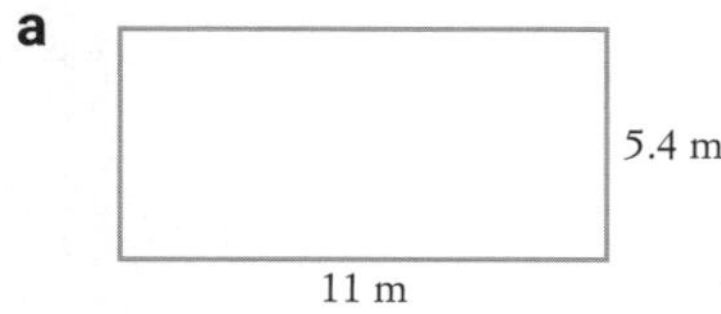

____

b

4 m

75 cm

____

5 Complete: $1\ m^2 =$ ________ cm^2.

HW HOMEWORK

PART C: PRACTICE

› Area of a triangle, parallelogram and composite shapes

HW HOMEWORK

1 Find the area of each shape.

a

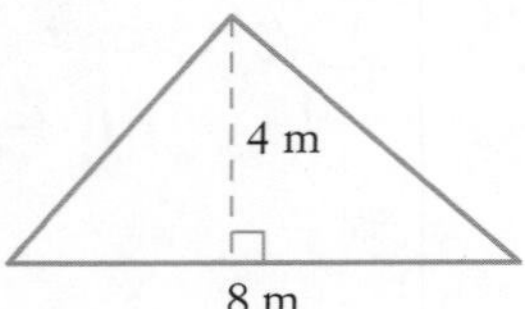

b

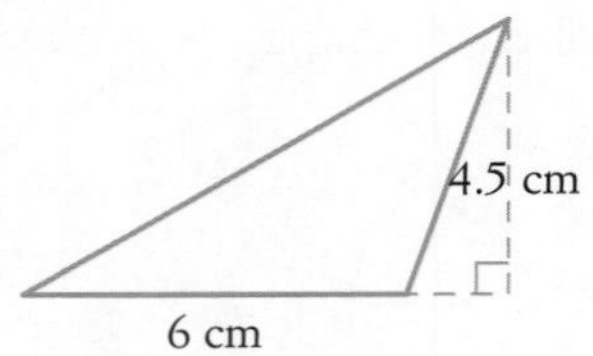

c

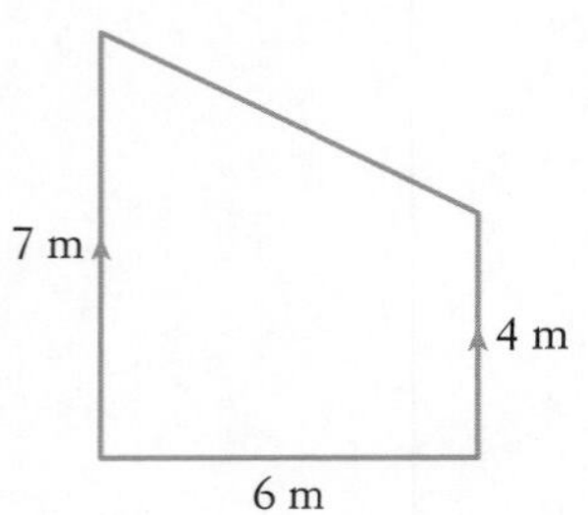

d

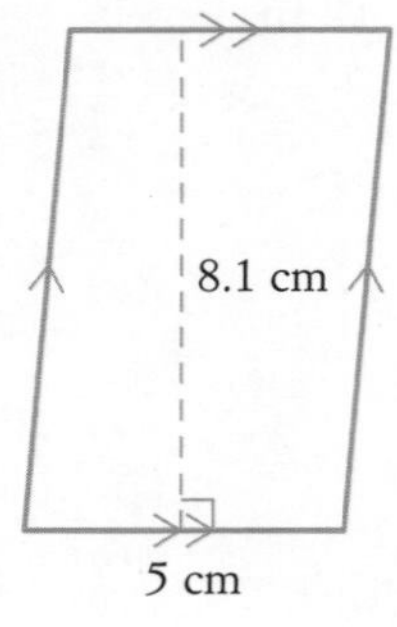

e

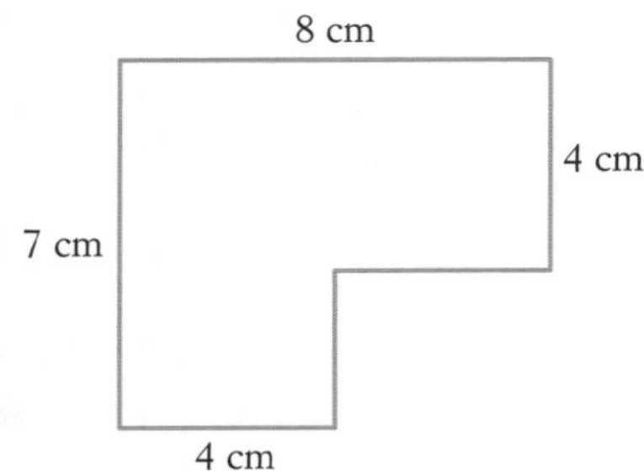

f

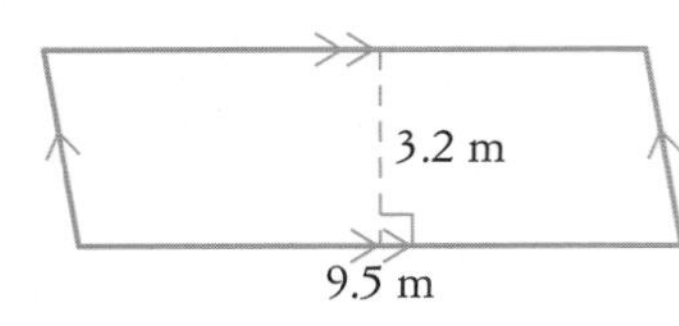

g

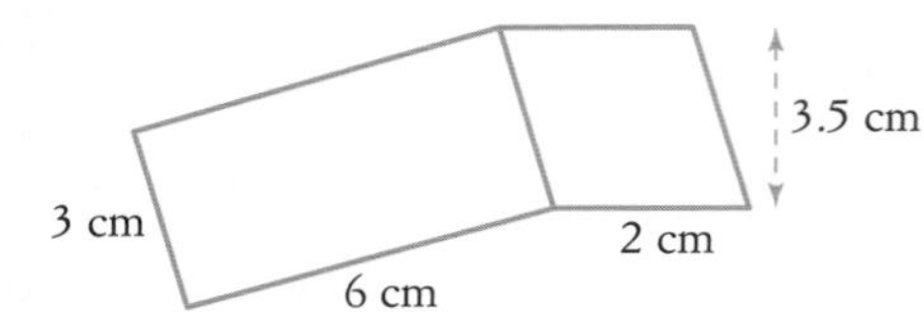

h

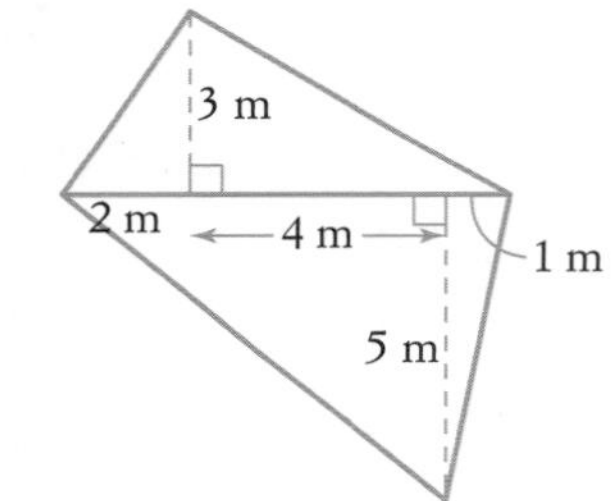

C S F

9780170454452

PART D: NUMERACY AND LITERACY

1 Which unit of area is equal to 10 000 m^2?

____

2 a What is the formula '$A = \frac{1}{2}bh$' used for?

____

____

b What does each variable in the formula stand for?

____

____

____

3 a Sketch an isosceles right-angled triangle with equal sides of length 5 cm.

b Find the area of this triangle.

____

4 a Write the formula for the area of a parallelogram.

____

b Describe in words how to find the area of a parallelogram.

____

____

5 Complete this triangle so that it has an area of 20 m^2.

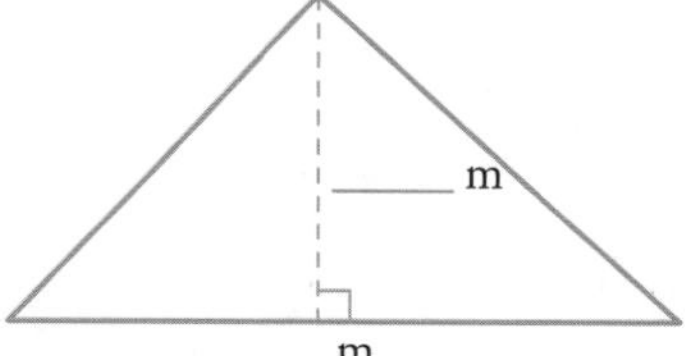

HW HOMEWORK

8 VOLUME AND CAPACITY

Name:

Due date:

Parent's signature:

Part A	/ 8 marks
Part B	/ 8 marks
Part C	/ 8 marks
Part D	/ 8 marks
Total	/ 32 marks

PART A: MENTAL MATHS

Calculators not allowed.

1 Evaluate 3^4. ____________

2 Complete this pattern:

4, 11, 18, 25, 32, ____________

3 If $a = 9$, evaluate $a^2 - 2a$.

____

4 Find n.

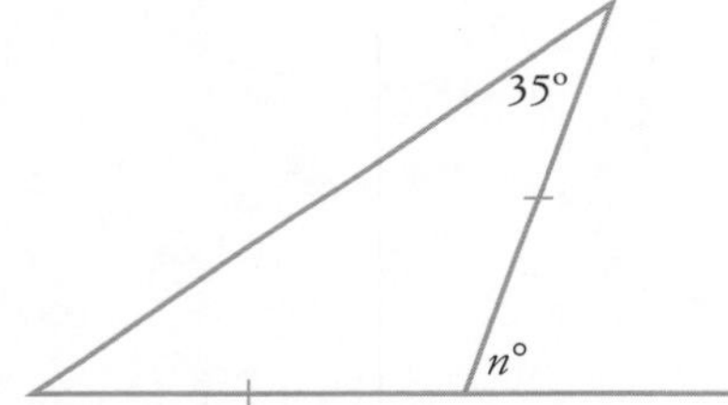

____

5 Evaluate $\frac{7}{10} \div \frac{4}{5}$.

____

6 Complete: 1 hour = ________ seconds.

7 Evaluate $-3 + 7 + (-1)$. ____________

8 Find the highest common factor of 16 and 20.

____

PART B: REVIEW

1 For this shape, find:

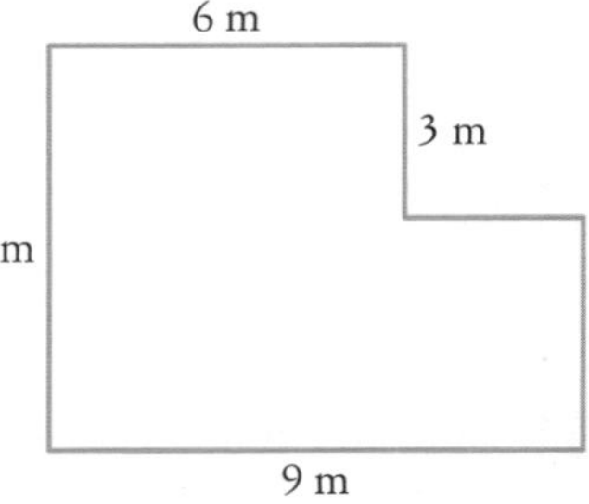

a its perimeter

____

b its area.

____

2 A rectangle has an area of 30 cm^2. If its width is 4 cm, what is its length?

____

3 Complete: 1 m^2 = ____________ cm^2.

4 Complete: 1 ha = ____________ m^2.

5 An equilateral triangle has a perimeter of 15 m. What is the length of each side?

____

6 Find the area of each shape.

a

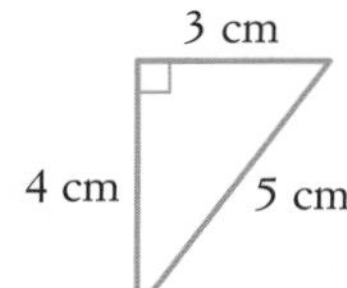

____

b

6 m

3 m

____

C S F

9780170454452

PART C: PRACTICE

› Volume of a rectangular prism
› Volume and capacity

1 Complete: 1 m^3 = ________ cm^3.

2 Complete: 1 mL = ________ L.

3 Find the volume of each prism.

a

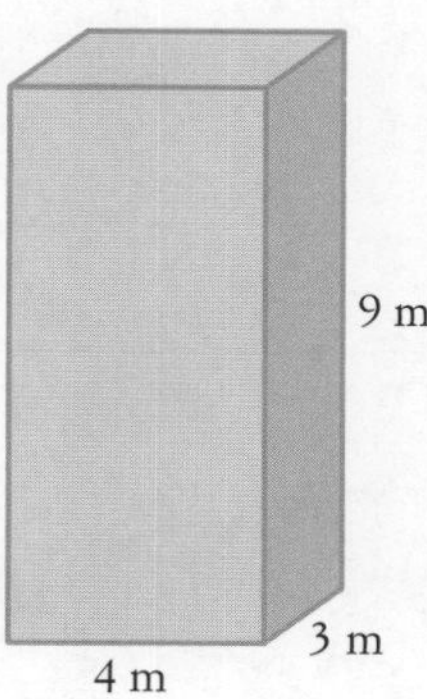

____

b

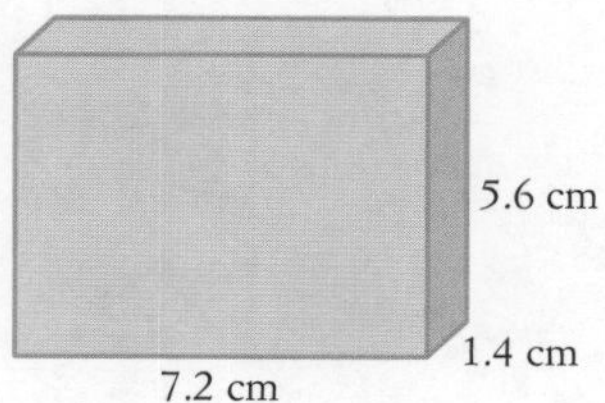

____

4 A cube has a volume of 64 m^3. What is the length of one side?

____

5 A rectangular pool has length 14 m and width 12 m. If it is filled to a depth of 3 m, how many kilolitres of water does it hold?

____

6 A rectangular box has length 75 cm, width 42 cm and height 30 cm. Find its volume:

a in cubic centimetres

____

b in cubic metres.

____

PART D: NUMERACY AND LITERACY

1 Describe a cubic centimetre.

____

____

2 Complete: If a solid has identical cross-sections that are polygons, then the solid is a

____

3 **a** Draw a square prism.

b What is a square prism?

____

____

4 What unit of capacity is equal to one cubic centimetre?

____

5 A rectangular garden is 6 m long and 1.5 m wide. It is filled with soil to a height of 38 cm. Find the volume of the soil in m^3.

____

6 **a** What is the formula $V = lwh$ used for?

____

____

b What does each variable stand for?

____

____

HW HOMEWORK

C S F

9 STARTUP ASSIGNMENT 9

LET'S GET READY FOR THE NUMBER PLANE TOPIC, WHICH COMBINES INTEGERS, ALGEBRA AND GEOMETRY.

PART A: BASIC SKILLS

/ 15 marks

1 Find d and e.

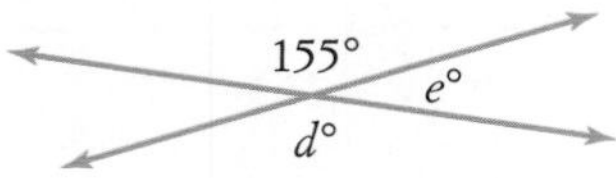

2 Round 17.8546 to 2 decimal places.

3 What time does a movie finish if it starts at 5:15 p.m. and runs for 108 minutes?

4 Evaluate:

a 45×8

b $\frac{3}{8} \div \frac{9}{16}$

c 3^4

d 25% of $260

5 When is the next year that is divisible by 9?

6 Simplify $6d - 5d$.

7 What are complementary angles?

8 Find x.

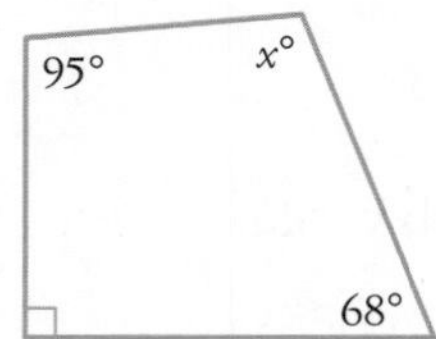

9 Write an expression for the number 4 more than p.

10 Write 0.45 as a simple fraction.

11 What is a square?

PART B: LOCATION AND TRANSFORMATION

/ 25 marks

12 For this map, what can be found at:

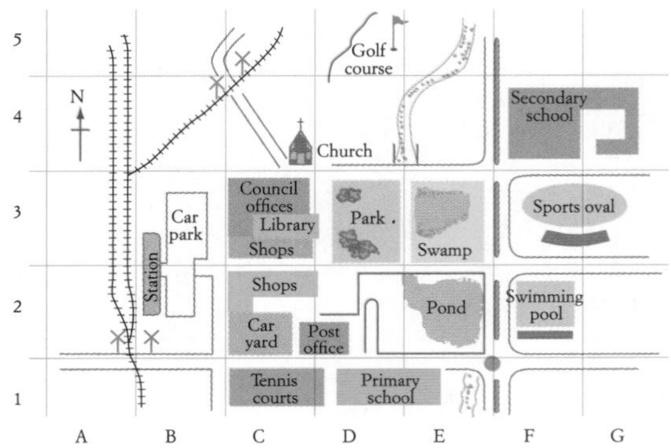

a F4?

b B2?

c D5?

13 For the above map, write the location of:

a the park

b the primary school

c the swimming pool

14 (2 marks) Complete this number line.

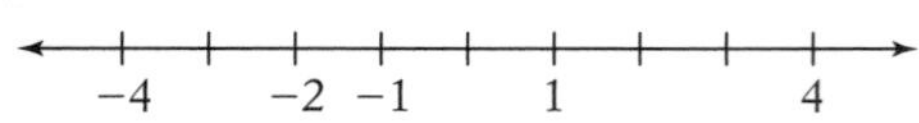

9780170454452

15 Translate *CBAD* 4 units right and 1 unit up.

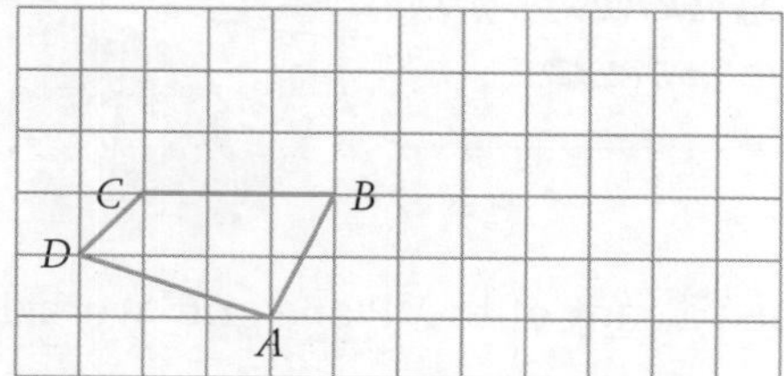

16 Rotate this trapezium 270° anti-clockwise about *D*.

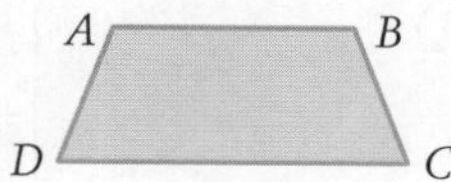

17 Reflect the figure across the dotted line.

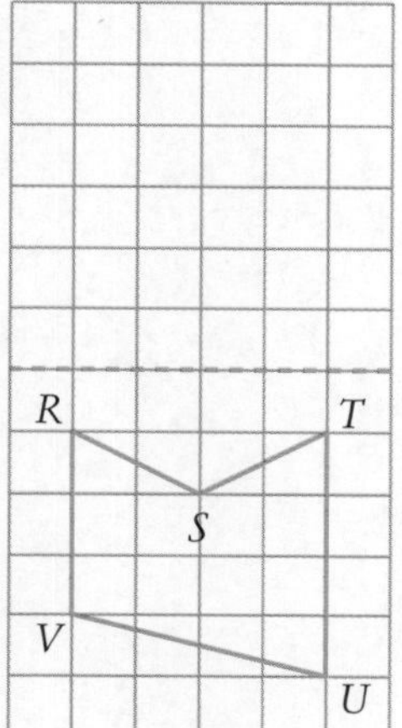

18 What is the size of one unit on this scale?

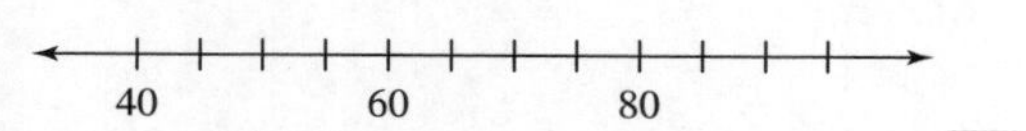

______

19 (4 marks) Write the coordinates of *A*, *B*, *C* and *D* on this number plane.

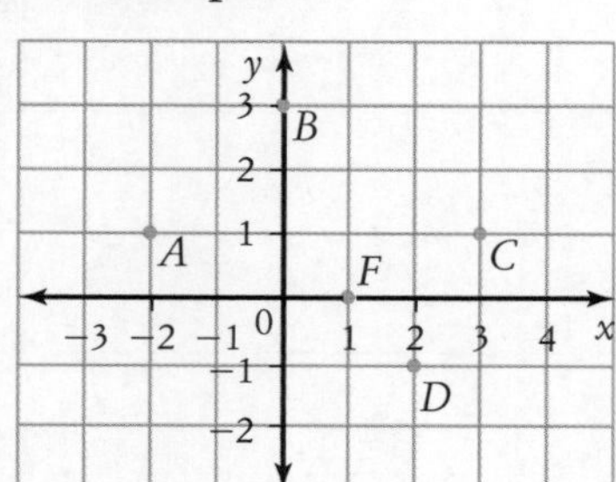

______

20 Which point on the above number plane lies on the *y*-axis? ______

21 (3 marks) On the number plane in question **19**, plot the points $E(-1, -2)$, $G(4, 1)$, $H(-3, 3)$.

22 Which 2 transformations were performed on figure *Q* to create image *Q′*?

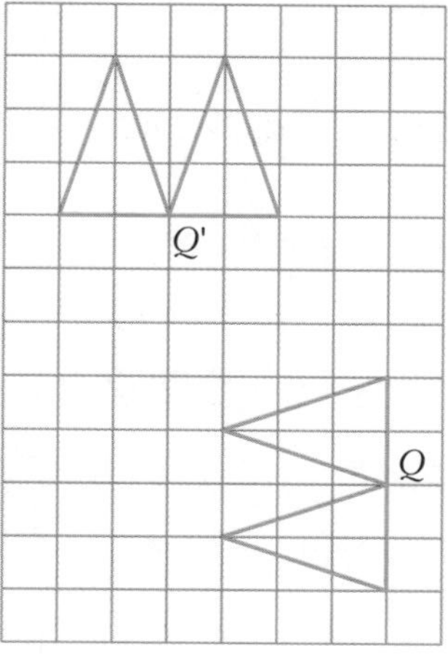

23 (2 marks) Which 2 points lie on the line?

$(3, -3)$, $(-3, -3)$, $(2, -1)$, $(1, 1)$, $(2, 7)$, $(-2, 1)$

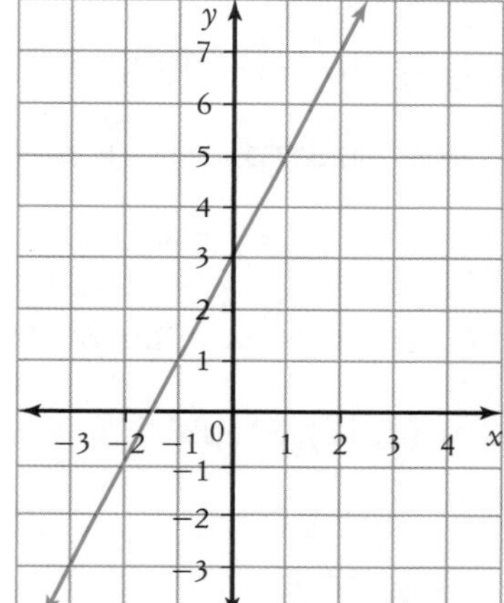

24 Complete: Another point that lies on the above line is (______, 5).

25 Write the coordinates of the point where the *x*- and *y*-axes cross. ______

PART C: CHALLENGE

Bonus / 3 marks

Ten years ago, Steve was as old as his wife, Linda, is now. Now Steve is twice as old as Linda was then. How old are Steve and Linda?

C
S
F

9780170454452

YOU'LL NEED A RULER AND YOUR COORDINATES SKILLS FOR THIS PICTURE PUZZLE.

Here are 2 number plane picture puzzles. For each puzzle, you need a copy of the 'Plane grid' worksheet (next page).

BIG TOP EMPLOYEE

1 Mark and join the following points in order.

- Join (1, 9), (3, 9), (3, 10), (5, 12), (8, 12), (10, 10), (11, 8), (10, 8), (11, 7), (10, 7), (11, 6), (10, 6), (10, 5), (9, 6), (10, 4), (9, 5), (9, 4), (8, 5), (8, 4), (7, 6), (7, 7), (8, 8), (8, 10), (7, 10), (6, 9), (6, 11), ($8\frac{1}{2}$, 11), **Stop**.
- Join ($7\frac{1}{2}$, 5), (7, 4), (8, 2), (1, 1), (7, 4), (2, 4), (3, 2) (1, 4), (1, 1), **Stop**.
- Join (5, 9), (4, 9), (4, 10), (5, 9), **Stop**.
- Join (3, 4), (2, 5), (2, 7), (3, 8), (1, 8), (1, 9), **Stop**.
- Join (3, 8), (5, 7), (4, 4), **Stop**.
- Join (5, 7), (3, 7), **Stop**.
- Join (5, 7), (4, 5), (2, 5), **Stop**.
- Join (5, 7), (4, 6), ($2\frac{1}{2}$, 6), **Stop**.
- Join (5, 10), (4, 11), **Stop**.

NAME THE SHAPE

2 Mark and join each set of points in order to form a shape. In each case, name the shape.

a (6, 12), (0, 6), (6, 0), (12, 6) ______________________

b (1, 10), (13, 14), (7, 19), (1, 17) ______________________

c (12, 4), (10, 8), (2, 4), (4, 0) ______________________

d (3, 9), (12, 12), (10, 18) ______________________

e (8, 20), (4, 16), (12, 16), (16, 20) ______________________

f (9, 12), (12, 6), (15, 12), (12, 18) ______________________

PLANE GRID

A NUMBER PLANE (GRID) FOR COORDINATE PUZZLES.

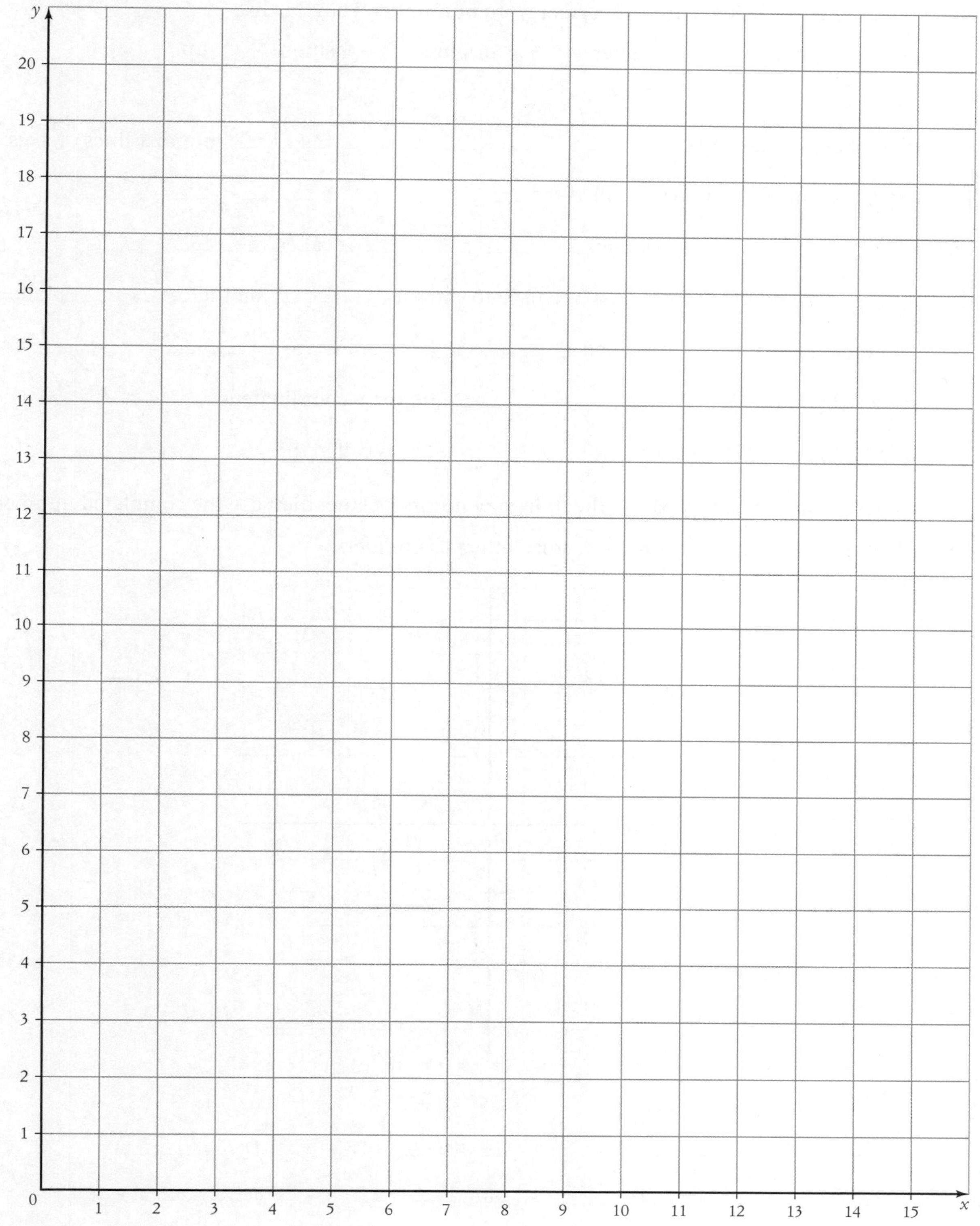

PS PUZZLE SHEET

y

20
19
18
17
16
15
14
13
12
11
10
9
8
7
6
5
4
3
2
1
0

1 2 3 4 5 6 7 8 9 10 11 12 13 14 15 x

9 THE NUMBER PLANE

THIS SHEET EXPLAINS AND REVISES THE NUMBER PLANE.

1 Complete the sentences below with the correct word or number from this list:

coordinates	cross	number	origin	position	up
vertical	*x*-axis	*x*-coordinate	*y*-axis	2	5

a A number plane is made up of ______ number lines or axes.

b The horizontal line going across is called the ______.

c The ______ line going ______ is called the ______.

d Two numbers written in the form (x, y) are used to show a ______ on the ______ plane.

e The two numbers in brackets are called ______.

f In the ordered pair (2, 5), the ______ is 2 and the *y*-coordinate is ______.

g The point (0, 0), where the 2 axes ______ is called the ______.

2 Fill in the missing values on the axes of the following number plane, then use the completed number plane to fill in the blanks below with the correct letters or numbers.

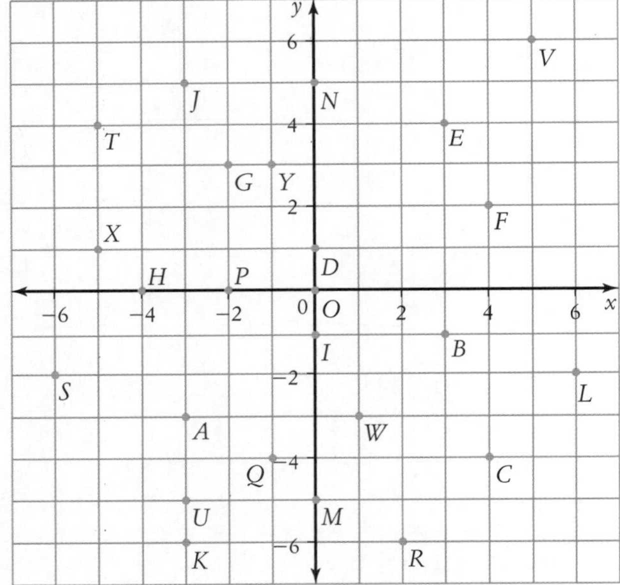

C(4, ______)	*H*(______, 0)	*N*(______, 5)	______(−3, −3)
T(______, ______)	*G*(−2, ______)	______(−1, −4)	*D*(______, ______)
K(______, −6)	______(−1, 3)	*M*(______, −5)	*J*(______, ______)
______(3, −1)	*P*(−2, ______)	______(−6, ______)	*U*(______, −5)
______(3, 4)	*I*(______, ______)	*V*(______, 6)	______(6, ______)
F(______, 2)	______(2, ______)	*O*(0, ______)	*X*(______, ______)

9780170454452

NUMBER PLANE CROSSWORD 9

THE ANSWER TO 1 ACROSS ALSO MEANS A PLACE WHERE SOMETHING BEGAN.

PS PUZZLE SHEET

Across

1 The point (0, 0)

3 Points in this quadrant have coordinates that are both positive.

6 The plural of 'axis'

7 Another name for number plane is ____ plane.

8 The point (5, 0) lies on the x-__________.

10 The 'spin' transformation

12 and 14 A grid in which positions are described with x- and y-coordinates (2 words)

13 Points in this quadrant have coordinates that are both negative.

15 Points that are collinear all lie on the same ____.

Down

2 This transformation creates a mirror image.

4 The 'slide' transformation

5 One quarter of the number plane

9 For $(8, -3)$, -3 is called the y-__________.

11 A list of (x, y) coordinates can also be written in a table of __________________.

14 $(8, -3)$ is sometimes called an ordered ______.

9 THE NUMBER PLANE 1

COORDINATES ON A MAP DESCRIBE A SQUARE REGION BUT COORDINATES ON A NUMBER PLANE DESCRIBE A PRECISE POINT.

Name:

Due date:

Parent's signature:

Part A	/ 8 marks
Part B	/ 8 marks
Part C	/ 8 marks
Part D	/ 8 marks
Total	/ 32 marks

HW HOMEWORK

PART A: MENTAL MATHS

Calculators are not allowed.

1 Evaluate 3×5.6. ______

2 Solve $\frac{m}{5} + 4 = 11$.

______

______

3 Write $3 \times 3 \times 3 \times 3 \times 3$ in index notation.

______

4 **a** Draw a parallelogram.

b What order of rotational symmetry has a parallelogram? ______

5 Find the highest common factor of 18 and 27.

______

6 Write an algebraic expression for the number of months in k years.

______

7 Find the area of this shape.

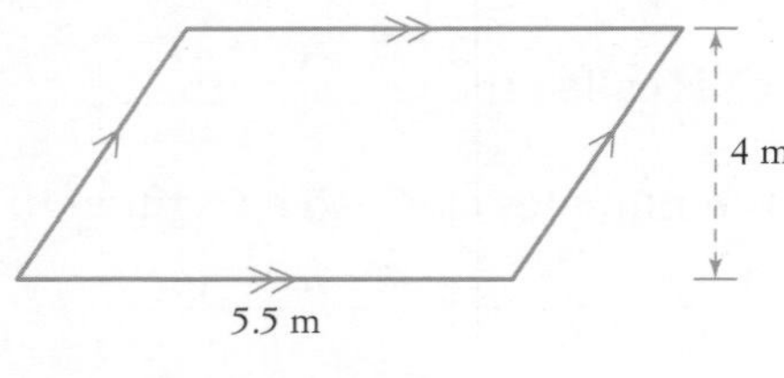

______

PART B: REVIEW

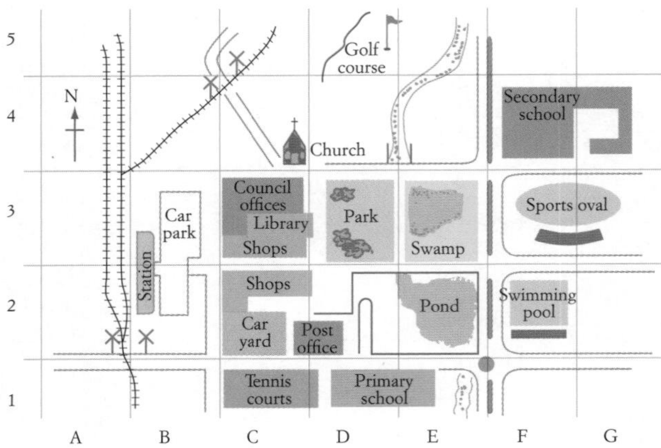

1 On the map, what is found at coordinates:

a G3? ______

b C4? ______

2 From the map, write the coordinates of:

a the pond ______

b the tennis courts ______

3 Write the coordinates of B on the number plane below. ______

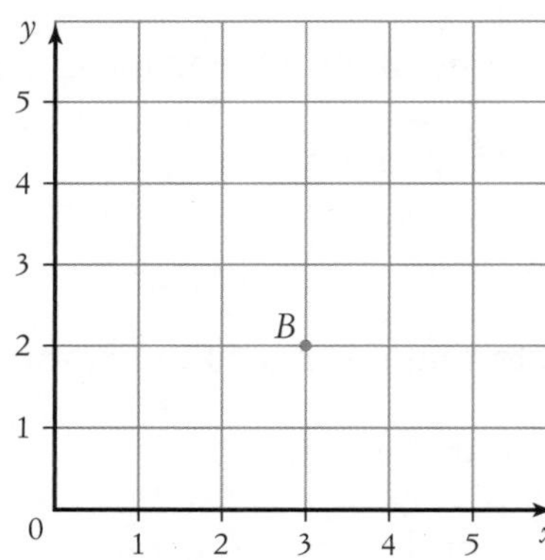

9780170454452

4 (2 marks) Plot and label $A(1, 4)$ and $C(4, 1)$ on the number plane above.

5 Write the coordinates of a point that is 3 units away from (2, 5). ____________

b What quadrilateral have you drawn?

____

c Name a pair of parallel sides.

____

PART C: PRACTICE

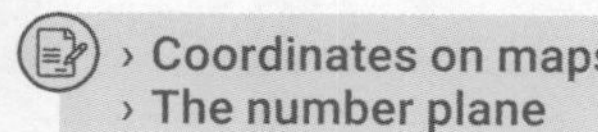

1 Use the number plane to write the coordinates of:

a P ____________

b Q ____________

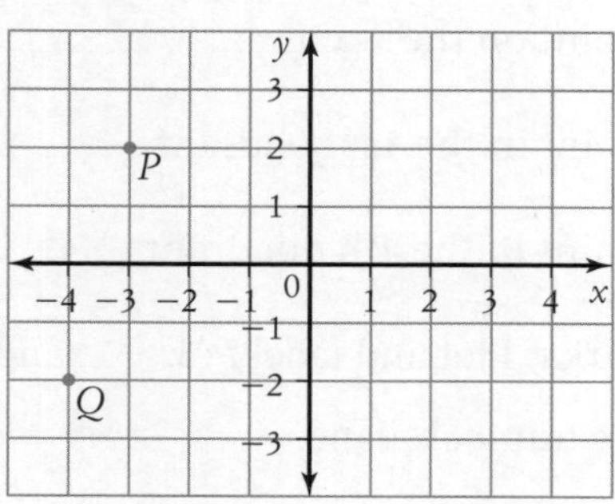

2 Which quadrant is point Q in? ____________

3 Plot and label $R(3, -2)$ on the above number plane.

4 a (2 marks) Plot and join these points on the number plane below: $W(-2, 2)$, $X(-2, -2)$, $Y(4, -2)$, $Z(2, 2)$, $W(-2, 2)$.

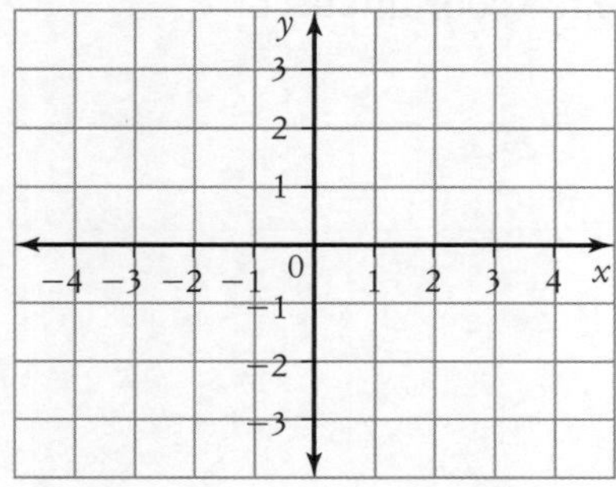

PART D: NUMERACY AND LITERACY

1 Other than a number plane, write an example of where grid coordinates are used.

____

2 Describe what each 'number plane' word means.

a y-axis

____

____

b origin

____

____

3 Complete: All points on the y-axis have an x-coordinate of ____________.

4 Complete: All points in the second quadrant have a ____________ x-coordinate and a ____________ y-coordinate.

5 (2 marks) Describe how to plot the point (2, −4) on a number plane.

____

____

____

____

6 What is another name for the horizontal axis?

____

C S F

HW HOMEWORK

9 THE NUMBER PLANE 2

THIS ASSIGNMENT SHOWS THAT NUMBER PATTERNS BECOME GEOMETRICAL PATTERS WHEN GRAPHED ON A NUMBER PLANE.

Name:

Due date:

Parent's signature:

Part A	/ 8 marks
Part B	/ 8 marks
Part C	/ 8 marks
Part D	/ 8 marks
Total	/ 32 marks

PART A: MENTAL MATHS

Calculators are not allowed.

1 Evaluate 0.08×0.4. ______

2 What shape is the cross-section of a cube?

______

3 Find the lowest common multiple of 8 and 18.

______

______

4 Evaluate $-1 + 3 - 4 - 2$. ______

5 Simplify $6 \times b \times a \times 2$. ______

6 Find the area of this shape.

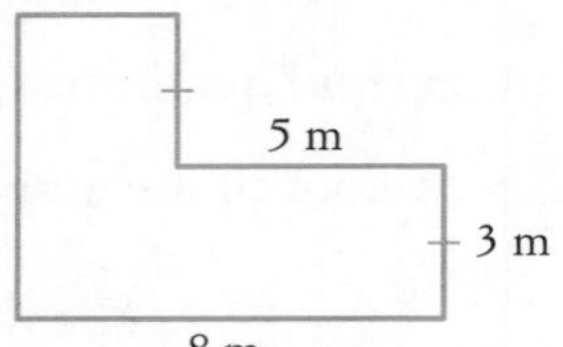

______

7 Evaluate 5^4.

______

8 Solve $3c - 8 = 13$.

______

______

______

PART B: REVIEW

1 Write the coordinates of:

a the origin ______

b a point on the x-axis ______

c a point in the 1st quadrant ______

d a point in the 4th quadrant ______

2 (2 marks) Plot and label $P(3, -1)$ and $Q(-4, -2)$ on the number plane.

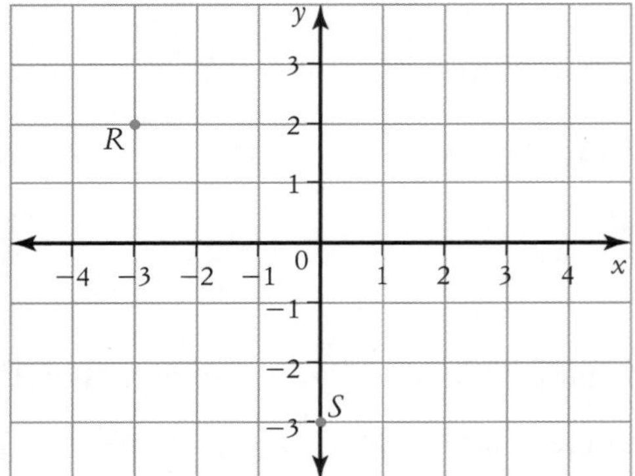

3 Write the coordinates of:

a R ______

b S ______

C S F

9780170454452

PART C: PRACTICE

› Graphing tables of values
› Transformations on the number plane

1 **a** (2 marks) Graph this table of values on the number plane.

x	−1	0	1	2
y	5	4	3	2

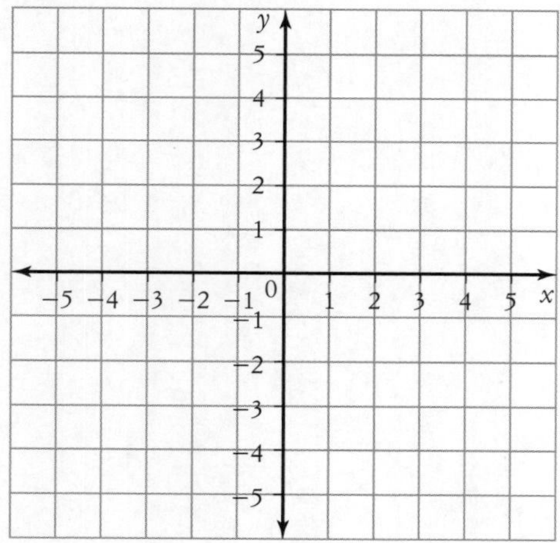

b What do you notice about the points?

---

c Write the coordinates of another point that follows the same pattern.

---

2 **a** What type of triangle is $\triangle LMN$?

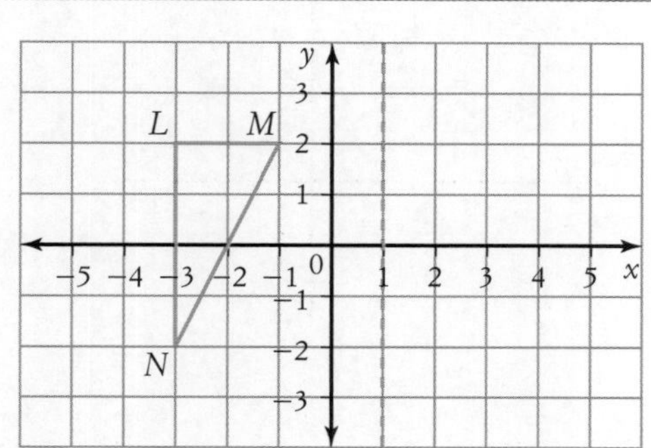

b Reflect $\triangle LMN$ across the dotted line to create $\triangle L'M'N'$.

C
S
F

c (2 marks) Write the coordinates of N and N'.

PART D: NUMERACY AND LITERACY

1 Accurately describe the transformation on shape P to produce P'.

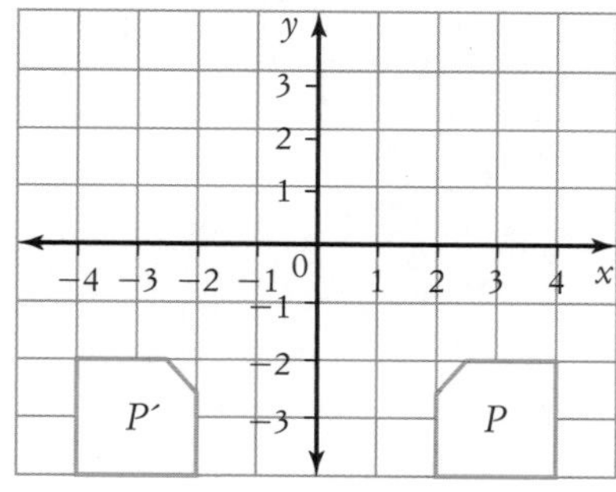

---

2 **a** Rotate $\triangle ABC$ 180° about $C(0, 0)$ to create $A'B'C'$.

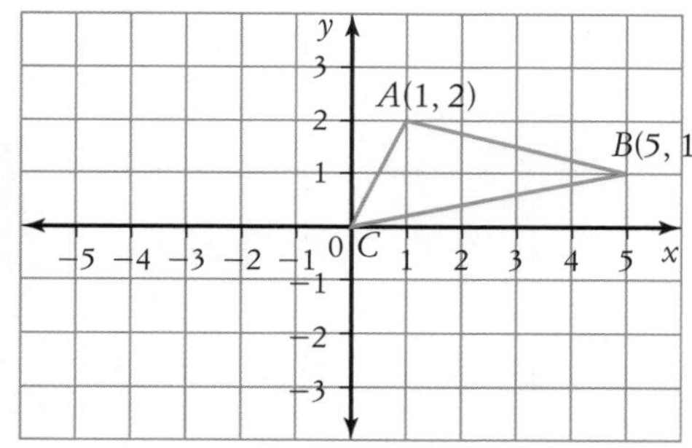

b (2 marks) Write the coordinates of A' and B'.

---

3 Complete: When a shape is reflected in the x-axis, the ________ -coordinate of every point in the shape changes sign.

HW HOMEWORK

4 **a** Translate the pentagon 4 units left and 2 units down to create $V'W'X'Y'Z'$.

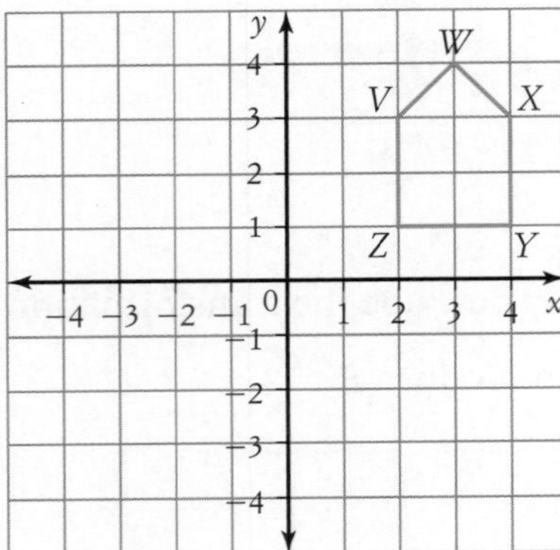

b (2 marks) Write the coordinates of Z' and describe how they are related to the coordinates of Z.

____

____

____

HW HOMEWORK

C S F

9780170454452

BEFORE WE START THE OUR DATA TOPIC, LET'S REVISE OUR STATISTICAL GRAPHS.

WS WORKSHEET

PART A: BASIC SKILLS

/ 15 marks

1 **a** What type of angles are shown on these parallel lines?

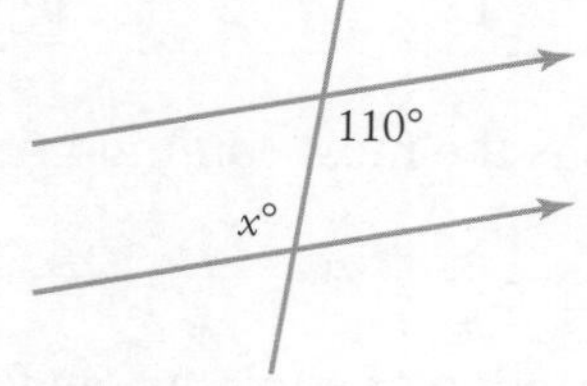

b Find x. ____________

2 Which is larger: 0.6 or $\frac{2}{3}$? ____________

3 What is the angle sum of a quadrilateral?

____

4 Evaluate:

a $-5 - (-11)$ ____________

b $\frac{2}{3} + \frac{1}{4}$ ____________

c 18×12 ____________

d 20% of \$450 ____________

5 Write an expression for the number that is 10 less than twice k. ____________

6 Simplify $3e \times 7d$. ____________

7 For this shape, find:

a the area

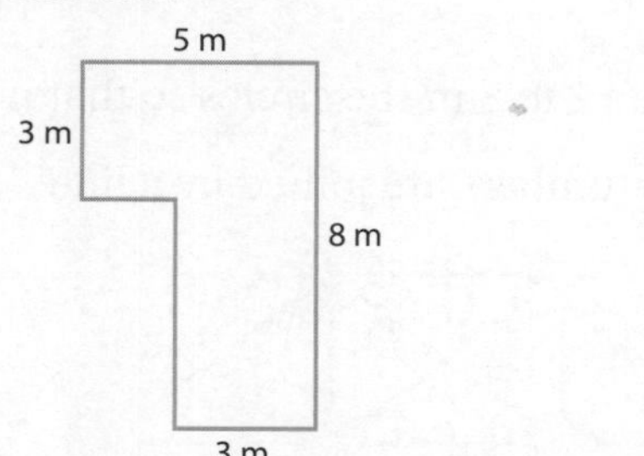

____

b the perimeter____________

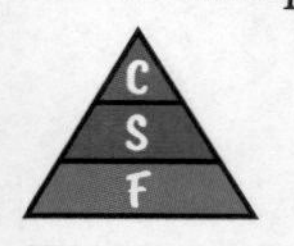

8 A tap leaks 20 mL of water each minute.

How much water will be lost in one hour:

a in mL? ____________

b in L? ____________

9 Which point on the number plane is closer to the x-axis: (7, 3) or (3, 7)?

____

PART B: GRAPHS AND DATA

/ 25 marks

10 Match each graph to its correct name: column graph, divided bar graph, line graph, sector graph.

a

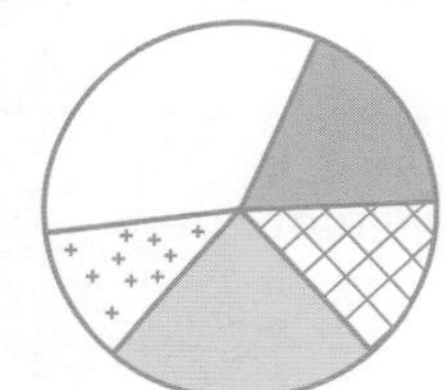

b

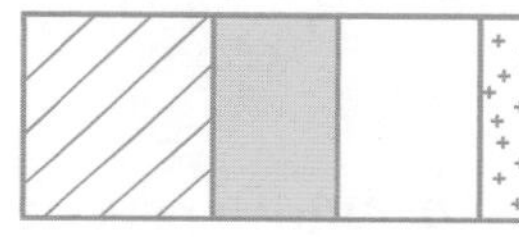

c

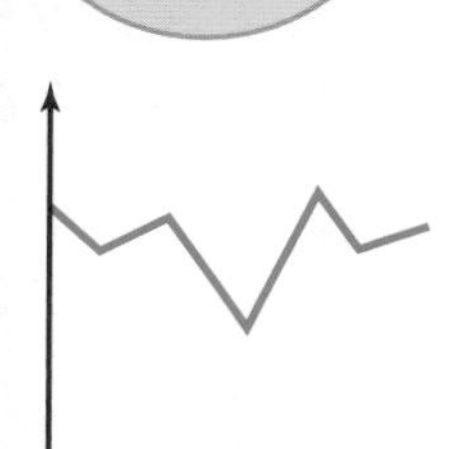

d

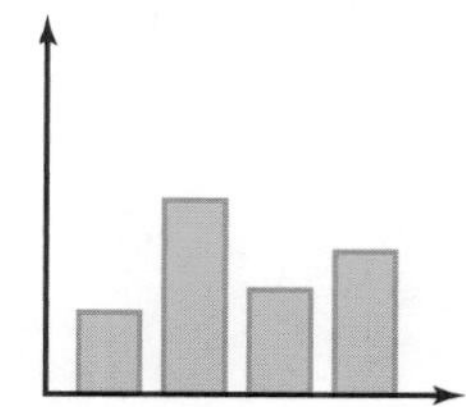

11 For this list of values:

6, 13, 7, 13, 10, 11, 16, 13, 10

a count the number of values ____________

b write them in ascending order

____

c write the most common value ____________

d write the middle value ____________

e find the difference between the highest and lowest values ____________

f find the average ____________

12 Complete each scale.

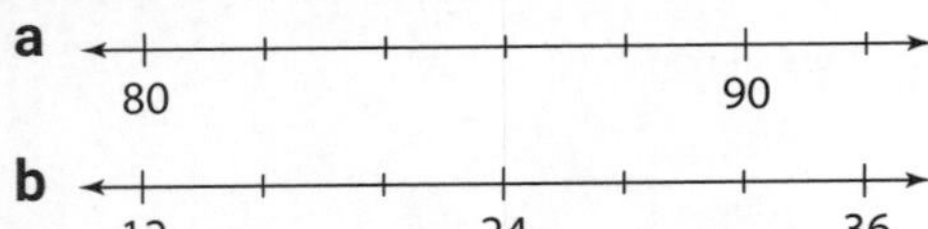

13 What fraction of this circle is shaded? ______

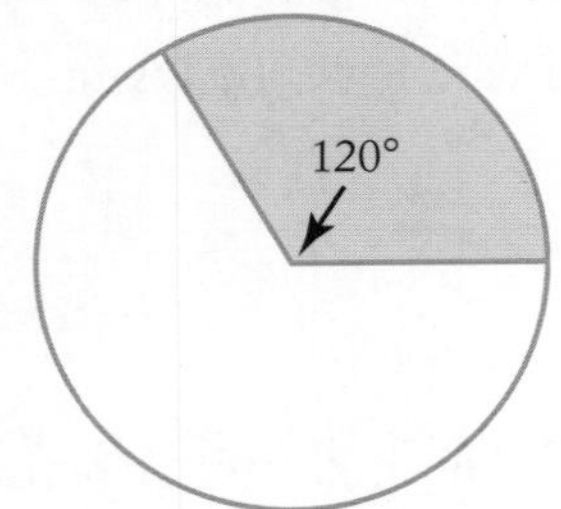

14 The results of a student survey on favourite pets are: Bird 20, Fish 25, Cat 50, Guinea pig 10, Dog 75, No pet 20.

a What was the least popular pet? ______

b How many students were surveyed? ______

c Which pet was the favourite for 25 students?

______

d Which pets were liked by more than $\frac{1}{5}$ of students? ______

e Illustrate the survey data on a column graph.

15 Find the average of 15 and 19. ______

16 Write 18 out of 24 as:

a a simple fraction ______

b a percentage ______

17 Find y if this sector is $\frac{1}{8}$ of a circle. ______

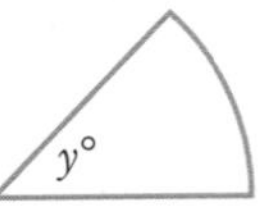

18 The results of a student survey on the number of TVs owned at home are:

1 TV	2 TVs	3 TVs	4 TVs	Over 4
10	12	3	0	2

a What was the most common number of TVs owned? ______

b What fraction of students owned 3 TVs?

______

c Illustrate the data using a picture graph, where 🯅 = 2 students.

PART C: CHALLENGE

Bonus / 3 marks

Place the numbers 1 to 8 in the circles so that no two consecutive numbers are joined by a line.

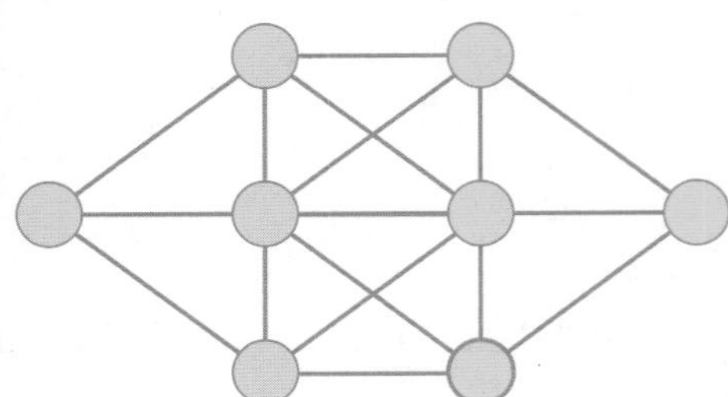

C S F

9780170454452

MEAN, MEDIAN, MODE 10

WS WORKSHEET

For each set of data, find the mean (to 2 decimal places where appropriate), median, mode(s) and range.

		Mean	Median	Mode(s)	Range
1	8, 11, 11, 15, 17, 20, 21				
2	10, 10, 16, 16, 20, 25, 26, 28				
3	3, 4, 5, 8, 11, 11, 14, 14, 14, 19, 20				
4	0, 1, 1, 1, 4, 5, 8, 10, 12				
5	18, 19, 20, 20, 20, 25, 27, 31, 34, 40				
6	7, 15, 19, 25, 29, 31, 40, 40, 50, 55				
7	54, 54, 60, 64, 64, 64, 68, 68, 70				
8	2, 1, 5, 3, 4, 5, 2				
9	91, 84, 86, 88, 90, 83				
10	10, 20, 30, 40, 50				
11	5, 4, 6, 3, 2, 4, 6, 9, 4, 7, 3, 2, 3				
12	16, 17, 19, 15, 17, 19, 14, 16, 17, 20				
13	7, 6, 9, 9, 8, 7, 7, 3				
14	17, 20, 19, 22, 21, 17, 100				
15	8, 10, 12, 7, 8, 10, 9, 8, 10, 8, 11				
16	21, 23, 20, 22, 21, 23, 24, 25				
17	47, 51, 48, 50, 48, 52, 51				
18	9, 9, 9, 9, 9, 10				
19	1, 0, 2, 3, 0, 1, 4, 2, 3, 0, 1, 1, 5, 4, 3, 2				
20	5, 9, 4, 6, 7, 8, 6, 5, 3, 2, 6, 4, 8				

(10) STEM-AND-LEAF PLOTS

STEM-AND-LEAF PLOTS ARE ONE WAY OF SHOWING DATA VALUES ORDERED AND GROUPED.

WS WORKSHEET

The heights of 28 students in a Year 7 class were measured (in cm).

182	166	166	172*	168*	161	175*
157	164	155	161*	173*	167	181
166*	168	165	165*	158*	163*	154
166	170*	181*	158	165	170*	170

(The 12 girls in the class are shown by *.)

These measurements can be grouped and listed in a **stem-and-leaf plot.**

Stem	Leaf
15	4 5 7 8 8
16	1 1 3 4 5 5 5 6 6 6 6 7 8 8
17	0 0 0 2 3 5
18	1 1 2

It groups the values into 'classes'; most heights are in the class '160–169'. An ordered stem-and-leaf plot is useful for organising a large number of data values. It is more powerful than a column graph for statistical calculations.

1 Use the stem-and-leaf plot to find:

a the modal height ____________

b the median height ____________

c the range ____________

d the mean (correct to 2 decimal places) ______

e how many students had a height of *over* 171 cm ____________

2 Suppose we wish to compare the heights of boys and girls. We can do this using a **back-to-back** stem-and-leaf plot. Complete the *ordered* plot below.

Girls		Boys
8	15	4 5 7 8
8 6 5 3 1	16	1 4
	17	
	18	

Note: Leaves branch outwards from the stem.

3 Just by looking at the distribution of data values, can you see which group:

a was taller, on average? ____________

b had a greater spread of heights? ____________

4 Estimate (guess) the answers to complete the following table.

	Girls	Boys
Median height		
Mean height		

9780170454452

5 Find the following measures (mean correct to one decimal place).

	Girls	Boys
Modal height		
Median height		
Mean height		
Range of height		

6 **a** How many boys were below 165 cm? ______

b How many girls were below 165 cm? ______

7 Write a brief report on what you have discovered about the differences between boys' and girls' heights in the Year 7 class.

8 Measure the heights of the students in your class and display the data in a back-to-back stem-and-leaf plot. Complete the table below and write a report on your conclusions.

	Girls	Boys
Modal height		
Median height		
Mean height		
Range of height		

WS WORKSHEET

10 STATISTICS 1

Name:

Due date:

Parent's signature:

Part A	/ 8 marks
Part B	/ 8 marks
Part C	/ 8 marks
Part D	/ 8 marks
Total	/ 32 marks

MATHS ISN'T JUST ABOUT NUMBERS AND ALGEBRA. IT'S ALSO ABOUT UNDERSTANDING STATISTICAL INFORMATION SHOWN ON GRAPHS AND TABLES.

PART A: MENTAL MATHS

Calculators are not allowed.

1 Simplify $7b - b$. ______

2 Evaluate $99 - 3 \times 8$. ______

3 Construct a 124° angle.

4 Write 8:28 p.m. in 24-hour time. ______

5 Is 23 583 divisible by 6? ______

6 Evaluate $0.024 \div 0.8$.

______

7 Write the angle sum of a quadrilateral.

______

8 Find the perimeter of this parallelogram.

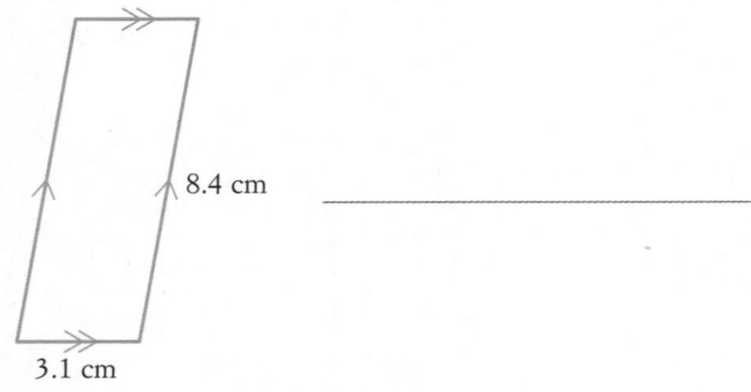

______

PART B: REVIEW

1 This graph shows the amount of TV watched by Zac over 5 days.

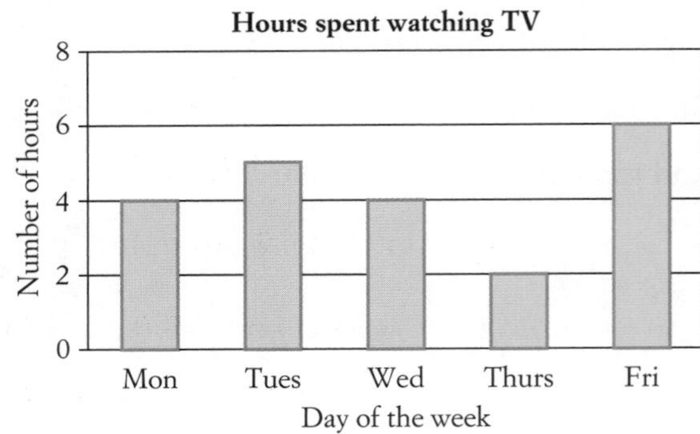

a What type of graph is this?

______

b What is shown on the horizontal axis?

______

c What is the size of one unit on the vertical axis?

______

d (2 marks) On which day did Zac watch the most TV? What could be a reason for this?

______

e How many hours of TV did he watch on Wednesday?

______

f On which day did he watch 5 hours of TV?

______

g Calculate the average number of hours of TV watched each day.

______

C S F

9780170454452

PART C: PRACTICE

› Interpreting graphs
› Misleading graphs
› Dot plots

1 This line graph shows the temperature of a town over 4 hours.

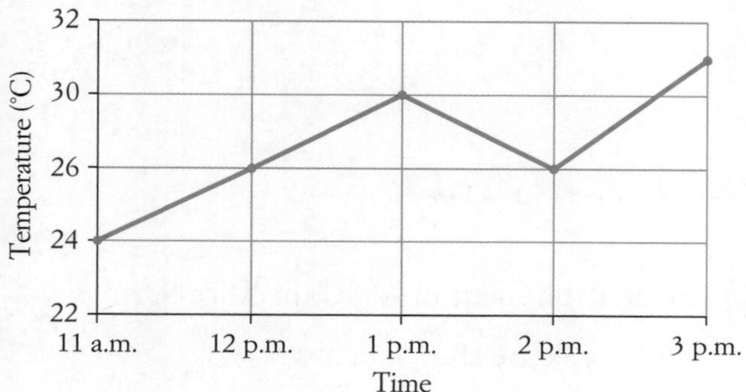

a Name one incorrect feature of this graph.

____

____

b What was the temperature at 1 pm?

____

c When was the temperature the highest?

____

d When was the temperature 24°C?

____

2 **a** Show these data values on a dot plot.

8 5 9 5 3 7 8 4 8 6 8 2

b What is the highest value? ______________

c What is the most common value? ______________

d What fraction of the values were over 5?

____

C
S
F

PART D: NUMERACY AND LITERACY

1 This graph shows how Helena spends a typical day, in hours per activity.

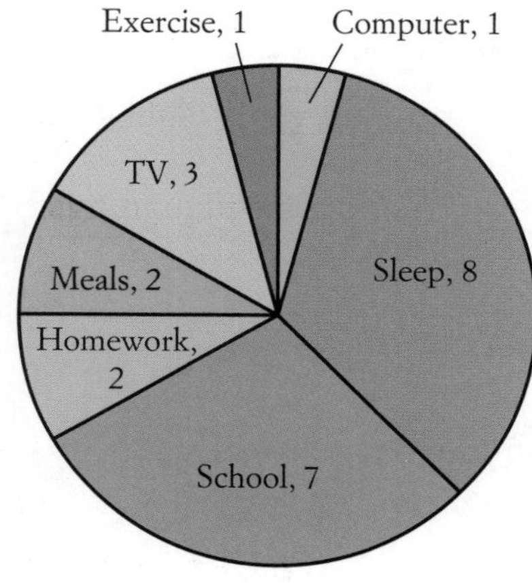

a What type of graph is this?

____

b Which activity takes up the most time?

____

c What fraction of the day is spent on TV?

____

d Which 2 activities take up 2 hours each?

____

e How many degrees are there in a revolution? ______________

____

f Calculate the sector angle for 'School', correct to the nearest degree.

____

g Name 3 activities that together make up $\frac{1}{4}$ of the day.

____

h Name one advantage of using this type of graph.

____

HW HOMEWORK

(10) STATISTICS 2

A SET OF DATA HAS ONLY ONE MEAN AND ONE MEDIAN, BUT IT CAN HAVE MORE THAN ONE MODE BECAUSE 2 OR MORE VALUES CAN BE EQUALLY MOST POPULAR.

Name:

Due date:

Parent's signature:

Part A	/ 8 marks
Part B	/ 8 marks
Part C	/ 8 marks
Part D	/ 8 marks
Total	/ 32 marks

PART A: MENTAL MATHS

Calculators are not allowed.

1 Draw a hexagon.

2 Convert $\frac{1}{6}$ to a recurring decimal.

____

3 Evaluate $-6 \times (-5) + (-4)$. ______________________

4 Write an algebraic expression for the number that is 10 less than y. ______________________

5 Name a quadrilateral with 4 equal angles.

____

6 Use a factor tree to write 90 as a product of prime factors.

____

7 Evaluate $\frac{2}{5} \times \frac{3}{4}$. ______________________

8 Find p. ______________________

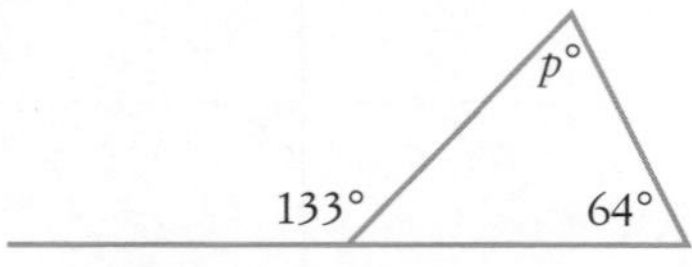

PART B: REVIEW

1 The daily temperatures (°C) in Alice Springs over a week are shown below.

Mon	Tue	Wed	Thu	Fri	Sat	Sun
22	20	22	18	17	16	15

a Write these temperatures in ascending order and find the middle temperature.

____

____

b Calculate the average temperature, correct to one decimal place.

____

c Draw a dot plot of the temperature data.

d What was the most common temperature?

____

e On what day was the temperature lowest?

____

f On how many days was the temperature above 18°C?

____

g Calculate the difference between the highest and lowest temperatures.

____

C S F

HW HOMEWORK

9780170454452

PART C: PRACTICE

› Stem-and-leaf plots
› The mean, mode, median and range

1 **a** How many data values are shown on this stem-and-leaf plot? ______________

Stem	Leaf
1	3 6 8
2	0 2 4 4 5 6 9
3	1 4 5 5 5 7 8 8
4	2 5

b What is the highest value? ______________

c What is the most common value? ______________

d What fraction of the values are below 30?

____

2 The ages of students in an art class are:

40 33 26 69 45 22 24 48 33 37 26 30 26

a Find the median age.

____

____

b What is the outlier? ______________

c Find the mean age correct to one decimal place.

____

d Draw an ordered stem-and-leaf plot for the ages.

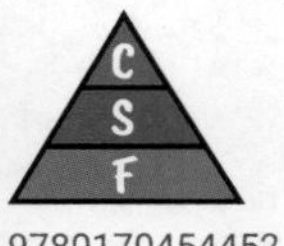

PART D: NUMERACY AND LITERACY

1 Name the statistical measure that is:

a calculated from the highest and lowest values of a data set

____

b the most popular value(s) of a data set

____

c calculated from *all* the values of a data set

____

2 What name is given to an extreme value in a data set that is much different to the other values?

____

3 The set of data below is in ascending order, with □ representing an unknown value.

3 3 4 8 □ 12

a Find the range.

____

b Find a value of □ if this set of data has 2 modes.

____

c Find a value of □ if this set of data has one mode.

____

d Find the value of □ if this set of data has a mean of 7.

____

____

____

HW HOMEWORK

10 STATISTICS 3

Name:

Due date:

Parent's signature:

Part A	/ 8 marks
Part B	/ 8 marks
Part C	/ 8 marks
Part D	/ 8 marks
Total	/ 32 marks

HW HOMEWORK

PART A: MENTAL MATHS

Calculators are not allowed.

1 Draw a rhombus and mark its axes of symmetry.

2 Find 25% of \$76.

3 Evaluate $7 \times 2 \times 50$.

4 True or false? For any 2 numbers a and b, $a - b = b - a$.

5 Solve $2u - 20 = 14$.

6 Evaluate $193.2 \div 1000$.

7 Find the area of this triangle.

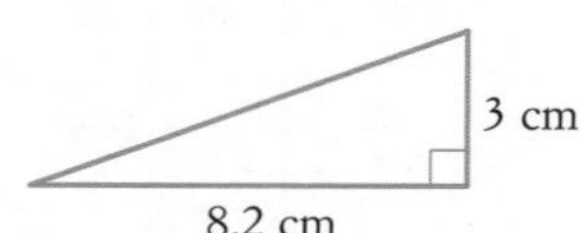

PART B: REVIEW

1 The heights (in cm) of a group of basketball players are:

179 185 204 183 192 188 170 185 195 199

a Find the median.

b Find the range.

c Which height is an outlier?

d Find the mean.

e Find the mode.

f Show these heights on an ordered stem-and-leaf plot.

g What fraction of players are over 190 cm?

h If the 192 cm player was replaced by a 182 cm player, would the mean increase or decrease?

C S F

9780170454452

PART C: PRACTICE

› Analysing dot plots and stem-and-leaf plots
› Comparing data sets

1 This dot plot shows the number of rainy days per week in Westvale over winter.

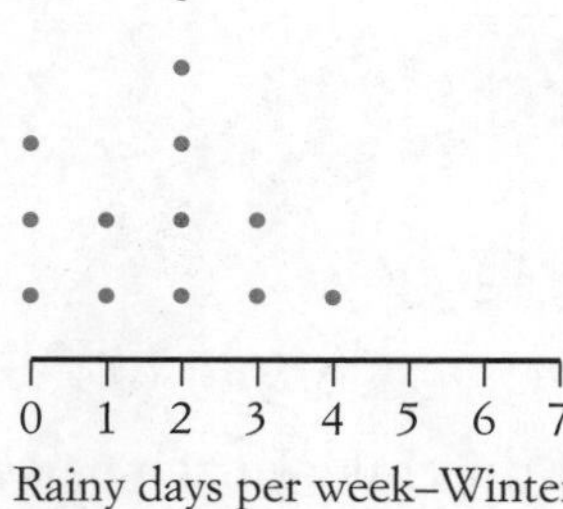

a Find the mode. ______________________

b How many weeks had only one rainy day?

____

c Find the range. ______________________

d Calculate the mean to one decimal place.

____

2 This dot plot shows the number of rainy days per week in Westvale over summer.

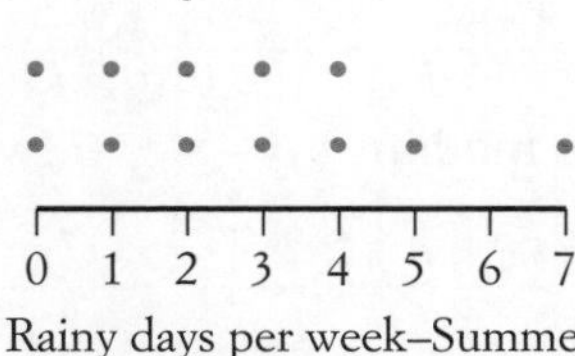

a What is the outlier? ______________________

b Find the median. ______________________

c Name one difference between the data for summer and winter.

____

d Which season has more consistent numbers of rainy days per week?

____

PART D: NUMERACY AND LITERACY

1 Name the statistical measure that:

a shows the spread of a data set

____

b can be more than one value

____

2 (2 marks) Name the 2 statistical measures that are most affected by outliers.

____

____

3 This stem-and-leaf plot shows the number of properties sold each month by a real estate agency.

Stem	Leaf
1	0 1 4 5 9
2	3 3 3 7
3	0 0

a Find the mode.

____

b Find the mean, correct to one decimal place.

____

c Find the median.

____

d Find the range.

____

HW HOMEWORK

C S F

10 STATISTICS REVISION

WE'RE GETTING CLOSE TO THE END OF THE DATA TOPIC NOW. IF YOU CAN SOLVE ALL THESE PROBLEMS, THEN YOU'RE A PRO AT STATISTICS.

Name:

Due date:

Parent's signature:

Part A	/ 8 marks
Part B	/ 8 marks
Part C	/ 8 marks
Part D	/ 8 marks
Total	/ 32 marks

HOMEWORK

PART A: MENTAL MATHS

Calculators are not allowed.

1 Evaluate $(-4)^2$. ______

______

2 Find $\frac{3}{8}$ of \$120.

______

3 Find x.

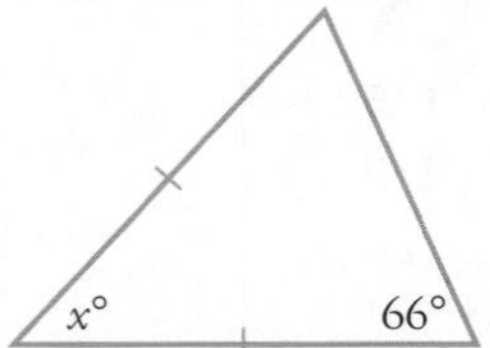

______

4 What is the time 2 hours 45 min after 3:20 p.m.?

______

5 List the factors of 20.

______

6 Draw a rectangular prism.

7 Find the lowest common multiple of 9 and 4.

______

8 The sum of twice a number and 12 is 40. What is the number?

______

PART B: REVIEW

1 The ages of the children at a party were:

5 6 3 6 7 5 4 5 9 5 6 7

a Represent these ages on a dot plot.

b What was the highest age?

______

c Find the mode.

______

d Find the median.

______

e Find the mean, correct to one decimal place.

______

f What fraction of children were aged 6?

______

g Find the range.

______

h If an 8-year-old joined this party, which measure would be most affected: the mean, mode, median or range?

______

C S F

9780170454452

PART C: PRACTICE

› Statistics revision

1 This stem-and-leaf plot shows the monthly costs in dollars of different business Internet plans.

Stem	Leaf
5	2 4
6	1 5 8
7	2 4 9 9
8	3 6
9	4

a What fraction of the plans had costs over $65 per month?

____

b Find the median.

____

c Find the mode.

____

d Find the mean.

____

e Find the range.

____

2 Suppose an outlier cost of $100 was added to the data in question **1**. Would each of the following measures be higher, lower or the same?

a The range

____

b The mode

____

c The mean

____

PART D: NUMERACY AND LITERACY

1 **a** Complete: This ____________ stem-and-leaf plot shows the Science project marks of 2 classes of students: 7W and 7S.

7W		7S
6	3	
	4	
8 2	5	
2	6	5 6
6	7	3 8
5	8	4 5 9
7 4	9	2 6 7

b What was the outlier mark? ______________

c Which class had a smaller spread of marks? ______________

d Comment on the difference between the marks of the 2 classes.

____

____

e Find the median mark for 7W.

____

f Find the median mark for 7S.

____

g Which class does not have a mode?

____

h Find the mean of 7W's marks, correct to one decimal place.

____

____

HW HOMEWORK

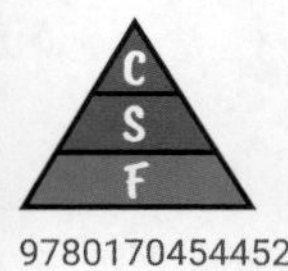

(10) DATA CROSSWORD

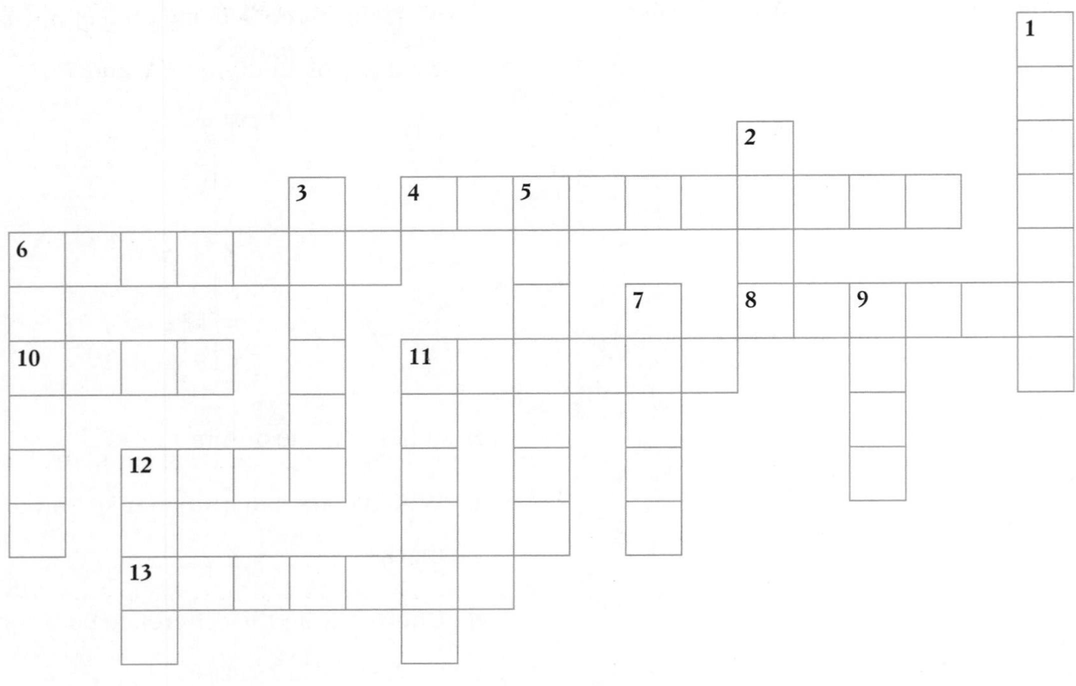

Across

4 The science of collecting and analysing data

6 A group of data values that are bunched or close together

8 The median is the _________ value.

10 In a stem-and-leaf plot, the unit digits of the values are listed in this column.

11 The range measures the _________ of the data.

12 Its symbol is $\bar{x}$

13 A simplified column graph that is easy to draw (2 words)

Down

1 An extreme data value

2 _________and-leaf plot

3 If there are an odd number of data values, then this is the middle value.

5 Another word for mean

6 Another name for bar chart is _________ graph.

7 The difference between the highest and lowest data values

9 A formal word for information

11 Another name for pie chart is _________ graph.

12 The most common value(s) in a set of data

PS PUZZLE SHEET

STARTUP ASSIGNMENT 11 (11)

THIS ASSIGNMENT REVISES SKILLS IN CHANCE AND NUMBER TO PREPARE YOU FOR PROBABILITY.

WS WORKSHEET

PART A: BASIC SKILLS

/ 15 marks

1 Evaluate:

a 6^3 ____________

b $\sqrt{121}$ ____________

c $\frac{2}{5} \times \frac{1}{4}$ ____________

2 Find x.

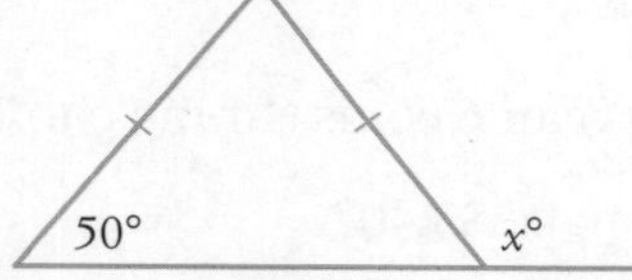

____

3 Solve $\frac{y + 4}{2} = 7$.

____

4 Write these decimals in descending order:

0.613, 0.6013, 0.6103, 0.6131

____

5 Complete each pattern:

a 23, 18, 13, 8, 3, ____________

b 1, 3, 6, 10, 15, ____________

6 What is the highest common factor of 32 and 28? ____________

7 What is the sum of cointerior angles on parallel lines? ____________

8 Find the volume of this cube.

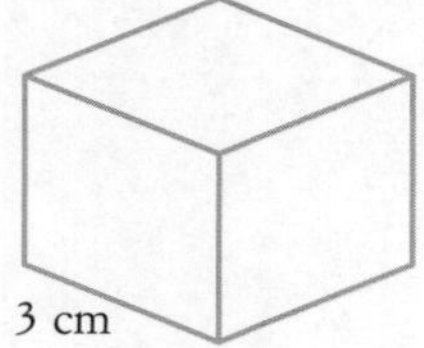

9 If $k = 4$ and $n = 1$, evaluate $5k - 2n$. ____________

10 A square has a perimeter of 36 cm. What is the length of one side? ____________

11 For this diagram, name a pair of:

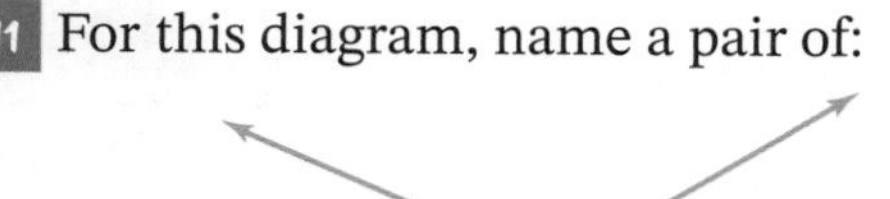
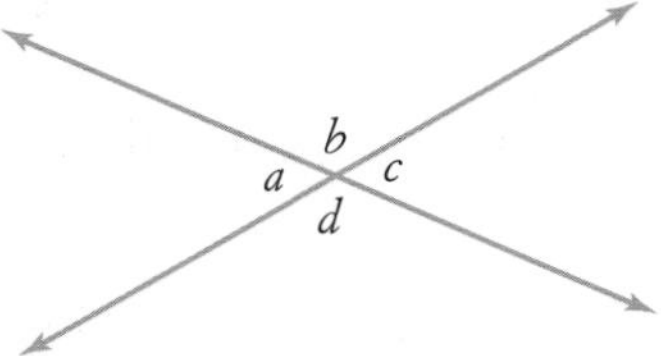

a supplementary angles ____________

b vertically opposite angles ____________

PART B: CHANCE AND NUMBER

/ 25 marks

12 Simplify:

a $\frac{12}{16}$ ____________

b $\frac{45}{80}$ ____________

c $\frac{18}{30} + \frac{7}{30}$ ____________

d $\frac{1}{4} \times 60$ ____________

13 How many:

a months begin with the letter A? ____________

b outcomes are possible when rolling a die?

____

C S F

WS WORKSHEET

14 If your soccer team plays a match, what are the possible outcomes of the match?

____

15 Convert to a percentage:

a $\frac{7}{8}$ ______________ **b** 0.05 ______________

16 Give an example of an event that has zero chance.

____

____

17 Simplify:

a $1 - \frac{1}{4}$ ______________

b $1 - \frac{7}{10}$ ______________

c $1 - 0.165$ ______________

18 What is the chance that the next baby born is a girl? ______________

19 Of the people at the Internet cafe, 68% was male. What percentage was female? ______________

20 Convert to a simple fraction:

a 35% ________ **b** 0.88 ________

21 (5 marks) Arrange these events in order, from the most likely to the least likely.

A Getting a tail when flipping a coin.

B A tennis player beating a slightly better player.

C You have something to drink later today.

D It snows at your house this month.

E You will be home by 4 p.m. today.

____

22 What are the possible outcomes for the colour displayed on a traffic light?

____

23 A bag contains 32 marbles, including 11 green, 6 red and 7 white marbles. The rest are blue.

a How many are blue? ______________

b If a marble is chosen from the bag without looking, what colour is it most likely to be?

____

PART C: CHALLENGE

Bonus / 3 marks

In how many ways can 6 coins (from 5c to $2 coins) be used to make $3.40?

____

C S F

9780170454452

GAMES OF CHANCE 11

LET'S LOOK AT THE CHANCES BEHIND A DECK OF CARDS, SPINNERS AND A ROULETTE WHEEL.

WS WORKSHEET

A DECK OF PLAYING CARDS

A full deck has 52 cards, comprising 4 suits of 13 cards each (Ace to King): hearts and diamonds (which are red), and clubs and spades (which are black).

1 A card is drawn at random from a normal deck of cards. Find the probability that it is:

a the 7 of hearts ____________

b a black card ____________

c a diamond ____________

d a Queen ____________

e a 3 or a 4 ____________

f a picture card ____________

g not a spade ____________

h an even number ____________

i a red card or an Ace ________

2 In poker, a player is dealt 5 cards. Brett is dealt 4 cards from a normal deck: 2 Aces and 2 Jacks.

What is the probability that the next card is:

a another Ace? ____________

b another Jack? ____________

c another Ace or Jack? ____________

d neither an Ace nor a Jack? ____________

3 Linda is dealt 4 cards — the 2, 3, 4 and 5 of hearts. What is the probability that the next card is:

a another 2, 3, 4 or 5? ____________

b another heart card? ____________

c an Ace or 6? ____________

d the Ace or 6 of hearts? ____________

4 Two cards are drawn at random from a deck of cards. The first card is shown to you.

a If the first card is a club, what is the probability of the second being a heart? ____________

b If the first card is a diamond, what is the probability of the second being another diamond? ____________

c If the first card is a King, what is the probability of the second being another King? ____________

d If the first card is the 10 of clubs, what is the probability of the second being the Jack of spades? ______

SPINNERS

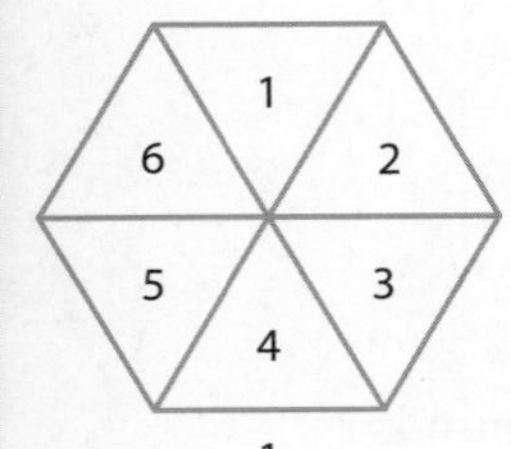

1

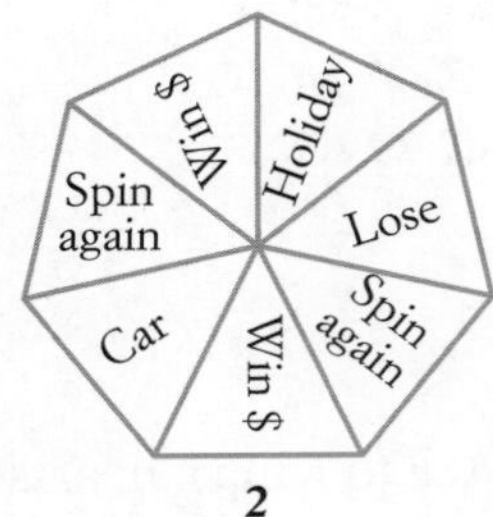

2

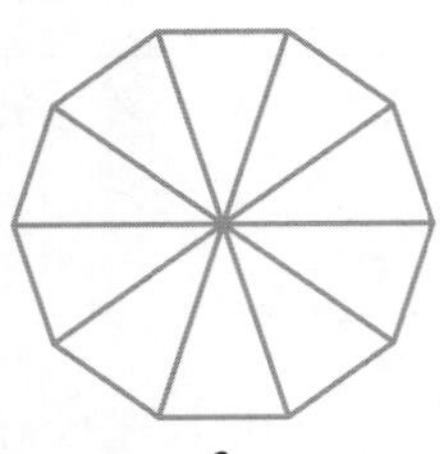
3

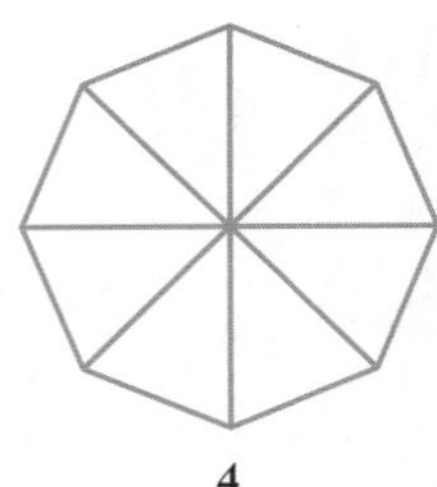
4

5 For spinner 1, what is the probability of spinning:

a a 3? ______

b a 7? ______

c an even number? ______

d a prime number? ______

e a factor of 12? ______

f a square number? ______

6 In one spin of spinner 2, what is the probability of:

a winning a car? ______

b winning money? ______

c losing? ______

d spinning again? ______

e winning anything? ______

f not winning money? ______

7 What angle is made at the centre by each section of spinner 3?

8 Design spinner 3 with prizes so that CASH has a 60% chance. Check your design by making it and spinning it 100 times.

9 Design spinner 4 with colours so that RED has double the chance of each of the other colours. Check your design by making it and spinning it 100 times.

ROULETTE WHEEL AND BETTING TABLE

A roulette wheel has 37 numbers from 0 to 36. Each number is coloured red or black, except for 0, which is coloured green.

		0		
1—18	1st 12	1	2	3
		4	5	6
EVEN		7	8	9
		10	11	12
◆ (red)	2nd 12	13	14	15
		16	17	18
◆ (black)		19	20	21
		22	23	24
ODD	3rd 12	25	26	27
		28	29	30
19—36		31	32	33
		34	35	36

10 For a roulette wheel, what is the probability of spinning:

a a 7? ____________

b a black number? ____________

c an even number (not 0)? ____________

d a number from 1 to 18? ____________

11 What is the probability of winning when betting on:

a a column of numbers? ____________

b a row of numbers? ____________

c the second 12? ____________

d 2 rows of numbers? ____________

e 4 numbers? ____________

f 2 columns of numbers? ____________

12 What is the probability that one of the numbers from 0 to 9 wins? ____________

13 Simone bets on 10, 20 and the third column. What is the probability that she wins? ____________

14 Rachel bets on 0, the high numbers (19 to 36) and the first 3 rows. What is the probability that she wins?

____

11 COINS PROBABILITY

YOU NEED 2 OR 3 COINS AND A CALCULATOR FOR THIS ACTIVITY. WORK IN GROUPS OF 2 OR 3.

1 When a pair of coins are tossed together, there are 4 possible outcomes. List them in this table.

1st coin	2nd coin	No. of heads
H	H	2
H	T	1

2 The possible number of heads can range from 0 to 2. Count the number of ways to get each number and complete this table.

No. of heads	Probability	Probability (%)
0	$\frac{1}{4}$	25
1		
2		

3 Why doesn't each event have the same chance?

____

4 Toss a pair of coins 80 times and count the number of heads each time. Tally your results in this frequency table.

Score	Tally	Frequency	Expected frequency
0			20
1			40
2			20
Totals		80	80

5 How were the expected frequencies calculated?

____

6 How do your results (observed frequencies) compare with the expected frequencies?

____

7 When 3 coins are tossed, there are 8 possible outcomes. Complete this table.

1st coin	2nd coin	3rd coin	No. of heads
H	H	H	3
H	H	T	2

 9780170454452

8 The possible number of heads now ranges from 0 to 3. Complete this table.

No. of heads	Probability	Probability (%)
0	$\frac{1}{8}$	12.5
1		
2		
3		

9 What number of heads has:

a the highest chance? ______________

b the lowest chance? ______________

Note the *symmetry* in the probabilities.

10 Toss 3 coins together 80 times and tally your results below.

Score	Tally	Frequency	Expected frequency
0			10
1			30
2			30
3			10
Totals		80	80

11 How do your observed frequencies compare with the expected frequencies?

____

12 The **law of averages** states that observed frequencies should become closer to the theoretical (expected) frequencies with more trials (tosses). Combine the results from all groups in your class and write them in the table below.

Score	Frequency	Expected frequency
0		
1		
2		
3		
Totals		

13 Are the observed frequencies closer to the theoretical frequencies now?

____

14 Construct a frequency histogram of these results. Is the graph symmetrical?

15 If 4 coins are tossed together:

a how many possible outcomes are there? ______

b what number of heads do you think will have the most chance? ______________

WS WORKSHEET

11 PROBABILITY 1

Name:

Due date:

Parent's signature:

Part A	/ 8 marks
Part B	/ 8 marks
Part C	/ 8 marks
Part D	/ 8 marks
Total	/ 32 marks

HW HOMEWORK

PART A: MENTAL MATHS

Calculators are not allowed.

1 Evaluate $1984 \div 8$. ______

2 Convert 0.3 to a percentage.

______

3 How many axes of symmetry has a scalene triangle? ______

4 Mark a pair of corresponding angles.

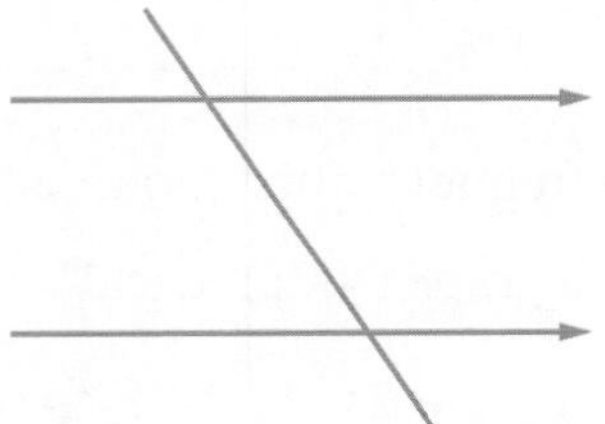

5 Convert $\frac{20}{6}$ to a simple mixed numeral.

______

6 Evaluate 7.85×10^3. ______

7 **a** Write an algebraic expression for 'triple a number x plus 6'.

______

b What is the value of x if triple a number x plus 6 equals 18?

______

______

PART B: REVIEW

1 Convert $\frac{17}{20}$ to a decimal. ______

2 How many:

a months begin with the letter J? ______

b prime numbers are below 20? ______

c Aces are in a deck of cards? ______

3 Select the correct word or phrase for each event: even chance, impossible, likely, unlikely.

a Selecting a red sock from a bag of white socks

______

b The next baby born being a boy

______

c Rolling 2 on a die

______

4 Name the 4 seasons of the year.

______

______

C S F

9780170454452

PART C: PRACTICE

› Sample spaces
› Probability

1 How many possible outcomes are there for:

a tossing a coin? ______

b a person's month of birth? ______

c a 2-digit number? ______

2 List the sample space for:

a the suit of a playing card

______

b the sex of a baby

______

3 A letter is selected at random from the alphabet. What is the probability, as a fraction, that it is a vowel or Y?

______

4 A box contains one of each Australian coin. If one is selected at random, what is the probability that it is:

a a gold coin?

______

b a coin whose value is a factor of 50c?

______

PART D: NUMERACY AND LITERACY

1 Write an example of a **certain** event.

______

______

2 Complete: In a probability situation, the set of all possible outcomes is called the ______

______.

3 A hospital has 12 male and 28 female nurses. One nurse is selected at random to visit a school.

a What does 'at random' mean?

______

______

b What is the probability, as a percentage, that the selected nurse is male?

______

4 Write the probability, as a fraction, for a 'fifty-fifty' chance. ______

5 What does '$P(E)$' mean?

______

______

6 **a** Write the 3 possible outcomes for the colour shown by a traffic light.

______

b Why is it incorrect to say that because there are 3 colours, then the probability that a traffic light shows red is $\frac{1}{3}$?

______

______

______

HW HOMEWORK

C S F

11 PROBABILITY 2

A PROBABILITY VALUE CAN BE WRITTEN AS A FRACTION, DECIMAL OR PERCENTAGE. THE PROBABILITY OF TOSSING TAILS ON A COIN IS $\frac{1}{2}$, 0.5 OR 50%.

Name:

Due date:

Parent's signature:

Part A	/ 8 marks
Part B	/ 8 marks
Part C	/ 8 marks
Part D	/ 8 marks
Total	/ 32 marks

PART A: MENTAL MATHS

Calculators are not allowed.

1 Is the point (0, 3) on the x-axis or y-axis?

____

2 Which quadrilateral has 2 pairs of equal adjacent sides?

____

3 Mark a pair of vertically opposite angles.

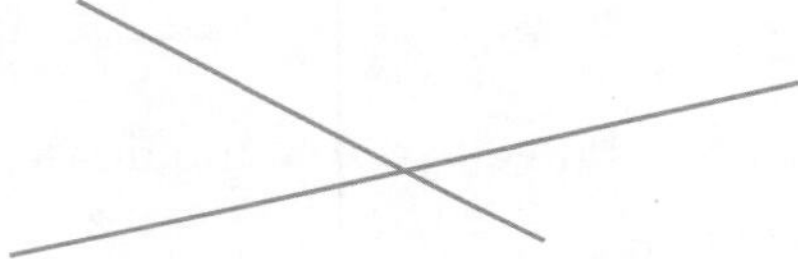

4 Round 9.8359 to the nearest tenth.

____

5 Evaluate $34.9 \div 10^4$. _______________

6 If $d = 2$, then evaluate $4 - 3d$.

____

7 Convert $\frac{1}{8}$ to a percentage.

____

8 Find the area of this triangle.

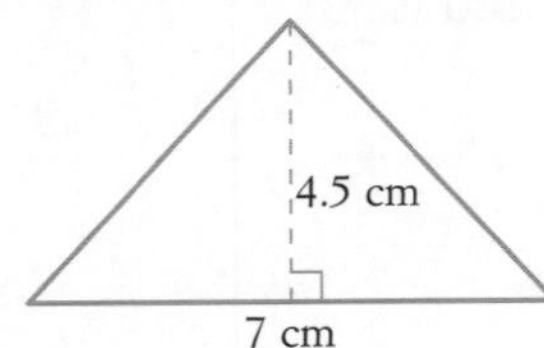

____

PART B: REVIEW

1 Evaluate $1 - \frac{27}{40}$. _______________

2 **a** List the sample space when a die is rolled.

____

b Find the probability of rolling a factor of 6 on a die. _______________

3 (2 marks) What is the difference between 'improbable' and 'impossible'?

____

____

____

____

____

4 A lolly is selected at random from a jar containing 12 red lollies and 4 yellow lollies. Find, as a decimal, the probability of selecting:

a a yellow lolly _______________

b a green lolly _______________

c a red or yellow lolly _______________

C S F

9780170454452

PART C: PRACTICE

› The range of probability
› Experimental probability
› Complementary events

1 Give an example of an event that has zero probability. ______

______

2 A box of chocolates contains 9 hard-centred and 11 soft-centred chocolates. One is chosen at random.

a Find the probability of selecting a hard-centred chocolate as a percentage.

______

b Find the probability of selecting a soft-centred chocolate as a percentage.

______

c Find the sum of the probabilities of selecting a hard- and a soft-centred chocolate, as a percentage.

______

3 How many heads would you expect if you tossed a coin:

a 30 times? ______

b 88 times? ______

4 A biased die was rolled 60 times, with the results shown in the table below.

Outcome	1	2	3	4	5	6
Frequency	15	8	2	10	8	17

a What is the relative frequency of rolling a 1? ______

b If this die is rolled 100 times, what is the expected frequency of rolling a 5? ______

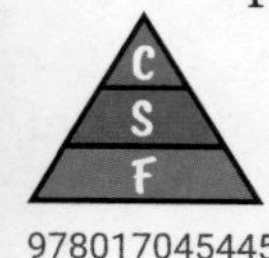

PART D: NUMERACY AND LITERACY

1 **a** What does 'even chance' mean?

______

______

b Give an example of an event that has an even chance.

______

______

2 What is the highest probability value?

______

3 The chance of rain on Thursday is 8%.

a Describe this chance.

______

______

b What is the probability that it won't rain on Thursday?

______

4 A magic show was performed to an audience of 5 11-year-olds, 7 12-year-olds and 8 13-year-olds. One person was selected at random from the audience. Find the probability of selecting a person whose age was:

a not 11

______

b not 10

______

5 What is the probability value of an impossible event?

______

HW HOMEWORK

11 PROBABILITY CROSSWORD

THIS CROSSWORD IS UNUSUAL BECAUSE THE ANSWERS ARE GIVEN BELOW. YOU JUST HAVE TO PUT THEM IN THE CORRECT PLACES!

The answers to this crossword puzzle are listed below. Arrange them in the correct places.

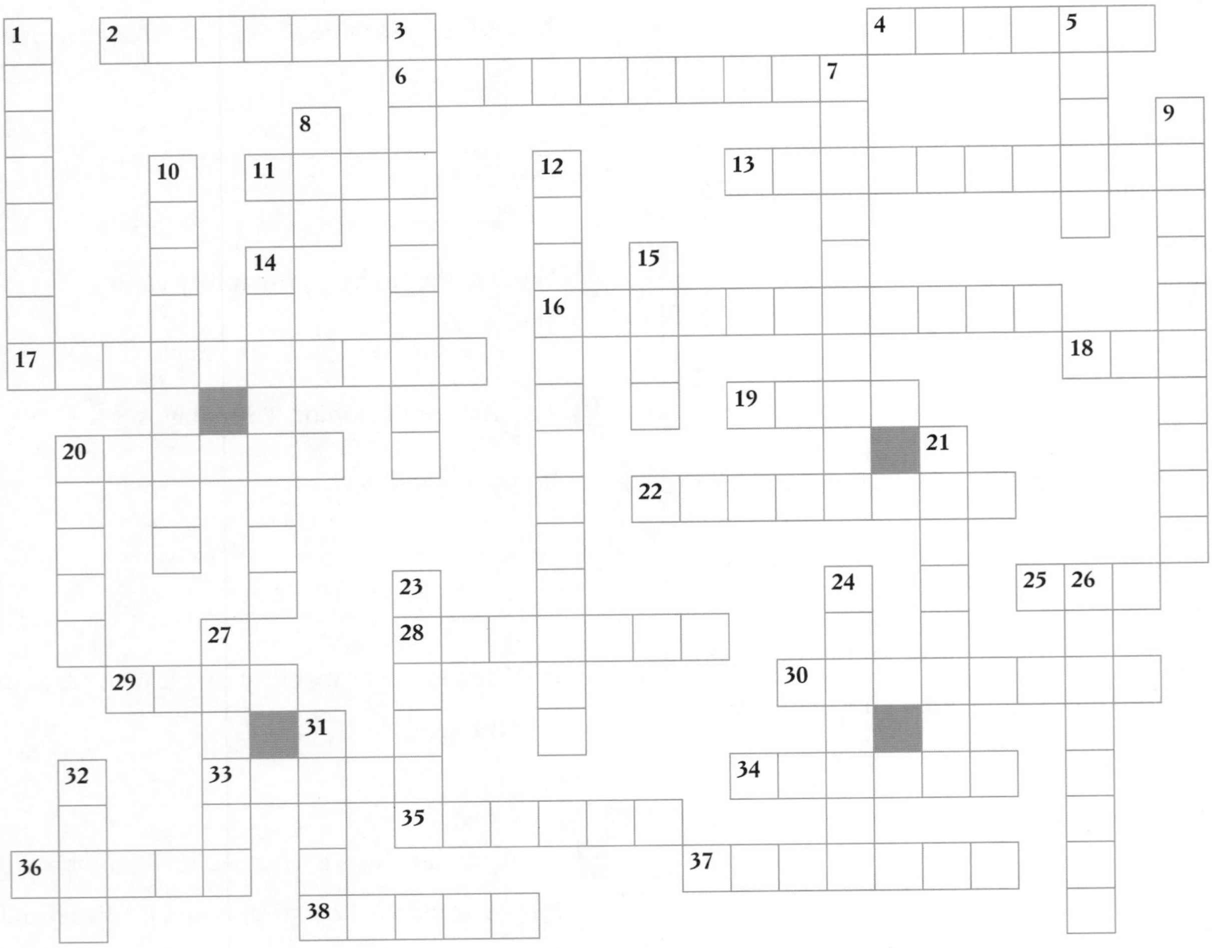

BUCKLEY'S	CERTAIN	CHANCE	CLUBS
COIN	COMPLEMENTARY	DECIMAL	DIAMONDS
DICE	DIE	EQUALLY	EXPERIMENT
EVEN	EVENT	FIFTY	FRACTION
FREQUENCY	HEADS	HEARTS	IMPOSSIBLE
IMPROBABLE	LIKELIHOOD	LIKELY	LOSS
PERCENTAGE	POSSIBLE	PROBABILITY	PROBABLE
RANDOM	RANGE	SAMPLE	SIMULATION
SPACE	SPADES	SPINNER	SUIT
SUM	UNLIKELY	WIN	

 9780170454452

STARTUP ASSIGNMENT 12 (12)

PART A: BASIC SKILLS

1 Write 0.325 as:

a a percentage ______

b a simple fraction ______

2 Does the point (8, 0) lie on the x-axis or y-axis of the number plane? ______

3 Simplify $4 \times n \div 3$. ______

4 Write a number between 750 and 760 that is divisible by 4. ______

5 Evaluate:

a $7.5 - 4.82$ ______

b $11.34 \div 0.7$ ______

c 32×5 ______

d $-5 + (-6) - (-7)$ ______

6 Classify this triangle:

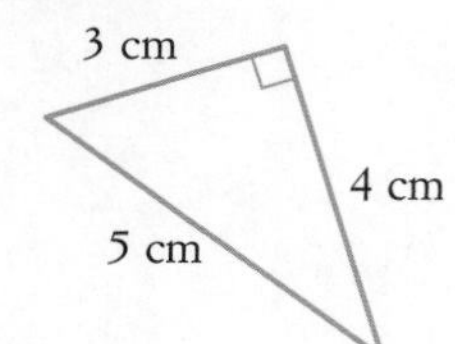

a by sides ______

b by angles ______

7 Find the area of the above triangle.

______

8 Which quadrilateral has one pair of parallel sides? ______

9 Find the mode of this set of data.

8, 1, 7, 9, 5, 8, 9, 1, 9. ______

10 Find x.

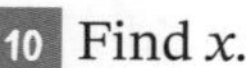

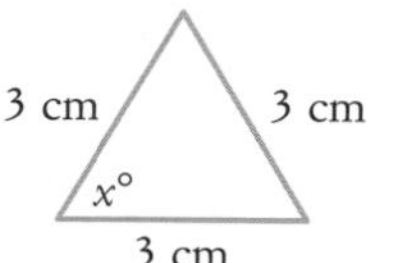

______

PART B: FRACTIONS AND TIME

/ 25 marks

11 What fraction is 20 g of 1 kg?

______

12 Write the time shown on this clock:

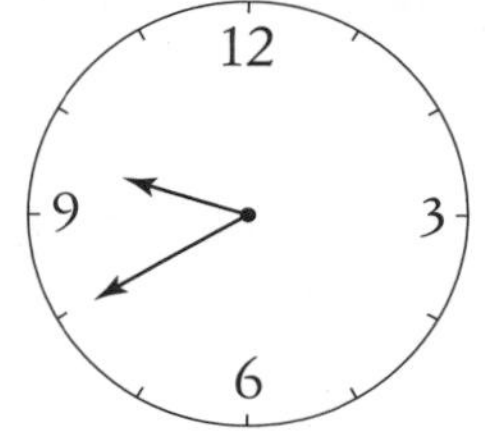

a in words

______

b in digital time

______

13 Write 4:50 p.m. in 24-hour time. ______

14 Write 21:25 in 12-hour time. ______

15 Simplify:

a $\frac{8}{40}$ ______ **b** $\frac{27}{45}$ ______

16 Complete:

a 3 hours = ______ minutes

b 5 days = ______ hours

c 2.4 L = ______ mL

d 8.5 m = ______ cm

C S F

WS WORKSHEET

WS WORKSHEET

17 What is the time:

a 5 hours after 1 p.m.? ____________

b 5 hours before 1 p.m.? ____________

18 The first modern Olympic Games were held in 1896 in Greece. How old are the Olympic Games this year? ____________

19 $\frac{2}{9}$ of the people at a party were adults. If there were 36 people at the party, how many were children? ____________

20 Complete:

a $\frac{2}{5} = \frac{10}{\square}$ **b** $\frac{3}{8} = \frac{\square}{48}$

21 Find the highest common factor of:

a 16 and 8 ________ **b** 30 and 42 ________

22 How many hours from 9 a.m. to 5 p.m.? ________

23 A jar has 13 yellow lollies, 18 green lollies and 20 red lollies. What fraction are green? ________

24 Find:

a $\frac{3}{5} \times 80$ ________

b $\frac{7}{9} \times \$342$ ________

25 Convert 450 minutes to hours and minutes.

____

26 If 32 L of petrol cost \$46.40, how much does one litre cost?

____

PART C: CHALLENGE

Bonus / 3 marks

Lindy wants to cook some pasta for 10 minutes but she doesn't have a watch or timer. Instead, she has a 7-minute hourglass and a 4-minute hourglass. How can Lindy use the 2 hourglasses to time 10 minutes?

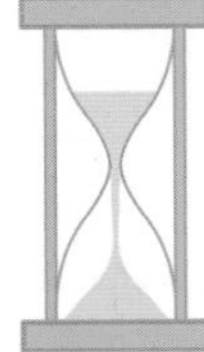
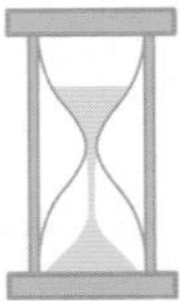

C
S
F

9780170454452

RATIOS AND RATES FIND-A-WORD 

HERE IS THE PUZZLE, HERE ARE THE WORDS. YOU KNOW WHAT TO DO!

T	V	S	S	L	E	P	H	J	X	M	U	K	C	W
T	N	P	M	E	K	X	F	Q	L	I	N	G	K	M
G	B	E	W	L	M	O	J	T	W	U	C	T	X	H
V	U	E	L	B	W	X	R	K	O	T	D	G	P	D
R	Y	D	S	A	S	A	D	Z	A	L	C	C	X	C
D	C	F	P	T	V	D	E	E	S	C	D	O	O	B
Y	A	R	N	E	H	I	E	Z	T	R	T	S	Y	I
U	I	R	L	M	W	S	U	Y	R	A	T	I	N	U
B	X	F	G	I	P	T	T	Q	T	E	R	M	S	S
G	S	C	X	T	C	A	J	C	E	N	P	P	S	H
B	M	E	Z	Y	H	N	H	P	A	R	G	L	O	X
S	E	M	A	T	H	C	N	O	T	V	L	I	B	J
U	K	W	B	P	P	E	U	G	N	N	T	F	I	D
P	J	W	K	H	S	A	Y	H	N	A	B	Y	W	T
V	G	K	E	S	T	C	A	X	R	U	O	H	T	U

Find these words in the puzzle above. They are across, up and down, and diagonal, and can be backwards as well as forwards.

BEST	BUY	COST	DISTANCE
EQUIVALENT	GRAPH	HOUR	PER
RATE	RATIO	SIMPLIFY	SPEED
STEEPNESS	TERM	TIMETABLE	TRAVEL
UNITARY			

PS PUZZLE SHEET

WS WORKSHEET

12 RATIO CALCULATIONS

RATIOS COMPARE PARTS OR SHARES OF SOMETHING. THIS SHEET HELPS YOU PRACTISE SIMPLIFYING RATIOS, FIND EQUIVALENT RATIOS AND SOLVE RATIO PROBLEMS.

1 Simplify each ratio.

a 10 : 75 ______

b 18 : 33 ______

c 8 : 64 ______

d 12 : 20 ______

e 27 : 18 ______

f 14 : 2 ______

g 36 : 8 ______

h 25 : 35 ______

i 40 : 10 ______

j 30 : 3 ______

k 16 : 48 ______

l 28 : 44 ______

m 25 : 30 : 50 ______

n 45 : 18 : 9 ______

o 122 : 18 : 28 ______

2 If 8 parts equal 72, what do 3 parts equal?

____

3 Which one of these ratios is equivalent to 4 : 7? Select **A**, **B**, **C** or **D**.

A 80 : 120 **B** 24 : 42

C 60 : 115 **D** 22 : 35

4 Simplify each ratio.

a 250 g to 1 kg ______

b 375 mL to 1 L ______

c 24 s to 1 min ______

d 4 cm to 2 m ______

e 1 day to 14 h ______

f 400 kg to 1 t ______

g 1200 m to 5 km ______

h 45 min to 1 h ______

i 2000 L to 10 kL ______

j 1800 cm to 1 m ______

k $4.00 to 78c ______

l 2.8 m to 560 mm ______

5 Which one of these ratios is equivalent to 8 : 3? Select **A**, **B**, **C** or **D**.

A 40 : 15 **B** 16 : 9

C 28 : 12 **D** 42 : 18

6 If 12 parts equal 180 min, what do 5 parts equal?

____

7 There are 36 male and 45 female teachers at Westvale High School. What is the ratio of male to female teachers? ______

8 Which one of these ratios is equivalent to 21 : 30? Select **A**, **B**, **C** or **D**.

A 3 : 10 **B** 7 :10

C 35 : 72 **D** 6 : 15

9 If 7 parts equal $147, what do 5 parts equal?

____

10 If 3 parts equal 735 mL, what do 11 parts equal?

____

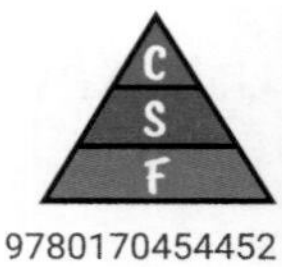

11 The ratio of tents to bushwalkers at the camp was 4 : 7. If there were 28 bushwalkers, how many tents were there?

____

12 Which one of these ratios is equivalent to 20 : 8? Select **A**, **B**, **C** or **D**.

A 6 : 5 **B** 15 : 6

C 5 : 1 **D** 4 : 2

13 Concrete mix should be made up of 2 parts cement and 5 parts sand. Find the amount of sand that needs to be mixed with 10 kg of cement.

____

14 The ratio of ice cream to milk in a milkshake should be 2 : 7. How much milk needs to be added to 150 mL of ice cream?

____

15 Two lengths of wood are in the ratio 4 : 7. If the longer piece is 35 cm, what is the length of the shorter piece?

____

16 The Directors of Centuryworld share its profits in the ratio 1 : 2 : 3. The Managing Director, who receives the greatest share, receives $9 144 000. What was the profit for that year?

____

C S F

WS WORKSHEET

12 RATE PROBLEMS

1 Write each of these as a rate:

- **a** 308 km in 3.5 hours = ________ km/h
- **b** $2.55 for 5 minutes = ________ cents/min
- **c** $6.80 for 8 kg = ________ cents/kg
- **d** 162 students with 3 teachers = ________ students/teacher
- **e** 81.9 m in 9 seconds = ________ m/s
- **f** $59.04 for 41 L = ________ cents/L
- **g** 696 words in 12 minutes = ________ words/min
- **h** 115 points in 5 games = ________ points/game

2 Write appropriate units for each rate.

- **a** population density ________
- **b** cost of fruit ________
- **c** fuel consumption ________
- **d** cost of a classified newspaper notice ________
- **e** download speed ________
- **f** cost of an information (1900) phone call ________

3 A 2.4 kg parcel costs $10.08 to send by post. What is the postage rate in $/kg?

____

4 A chef cooks 85 meals between 5 p.m. and 10 p.m. Calculate his average hourly rate of cooking.

____

5 The cost of water is $2.016/kL. A household uses an average of 850 L per day. Calculate:

- **a** the water usage over 6 weeks ________
- **b** the cost of water over 6 weeks ________

6 A petrol tanker discharges 680 L of fuel per minute. How long will it take to empty its tank of 8500 L?

____

7 Jade is driving at a speed of 81 km/h. How far will she travel in:

- **a** $4\frac{1}{2}$ hours? ________
- **b** 35 minutes? ________

8 Ilhea sheared 54 sheep in 4 hours. What was her shearing rate per hour?

____

9 Who earns the most per hour?

Baker: $297.30 for 8 hours

Hairdresser: $214.98 for 6 hours

Painter: $416.78 for 10 hours

Plumber: $577.20 for 12 hours

Roof tiler: $462.26 for 11 hours

____

10 Wollongong's population grew by 15 392 in 8 years. Calculate its annual growth rate.

____

11 Brett made 30 runs in 8 overs of cricket. Calculate his run rate in runs/over.

____

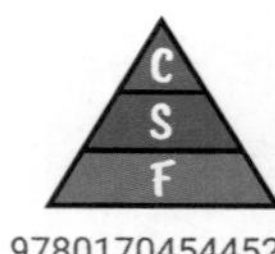

9780170454452

12 A water tank of 182 L capacity takes 35 minutes to drain. At what rate is the tank being drained?

___

13 A 13-minute mobile phone call costs $11.44. Find the rate charged:

a in cents per 30 seconds ___

b in cents per second, correct to one decimal place ___

14 Melinda earned $953.60 for working from 9 a.m. to 5 p.m. Monday to Friday. What is her hourly rate of pay?

___

15 A tap drips 28 litres of water each day.

a How many litres are wasted over 5 days?

___

b How many weeks will it take to waste 1176 L?

___

16 If Vicky's heart beats 72 times per minute, how long will it take to beat 1836 times?

___

17 Rasha swims at a speed of 1.78 m/s.

a How far will she swim in 5 minutes at this speed?

___

b How long will it take her to swim 400 m? Answer correct to the nearest second.

___

18 1020 kWh (kilowatt-hours) of electricity cost $292.64. What is this cost in cents/kWh, correct to 2 decimal places?

___

19 A satellite travelled 29 160 km in one hour. What was its speed in metres per second?

___

20 Petrol costs $1.51/L. How much does 43 L cost?

___

21 A mobile phone call costs 28c per 30 s. For how long could you talk for $30? Answer correct to the nearest minute.

___

22 In 2009, Jamaican athlete Usain Bolt ran a 100 m race at an average speed of 10.44 m/s. How many seconds, correct to 2 decimal places, did he take to run this distance?

___

WS WORKSHEET

12 TIME CALCULATIONS

1 Convert each time to 24-hour time.

a 6:30 a.m. ______ **b** 4:20 p.m. ______

c 11:05 p.m. ______ **d** 2:45 p.m. ______

e 7:18 p.m. ______ **f** 10:56 a.m. ______

2 Convert each time to 12-hour time.

a 07:05 ______ **b** 18:55 ______

c 12:30 ______ **d** 20:30 ______

e 03:44 ______ **f** 15:17 ______

3 What is the time:

a 6 hours after 3 p.m.? ______ **b** 5 hours after 8 a.m.? ______

c 14 hours after 17:00? ______ **d** 7 hours before 10 p.m.? ______

e 10 hours before 19:00? ______ **f** 16 hours before 12 midnight? ______

g 13 hours after 9:30 a.m.? ______ **h** 9 hours before 14:40? ______

i $8\frac{1}{2}$ hours after 6 p.m.? ______

4 Use the °′″ or D°M′S key on your calculator to convert each time to hours and minutes.

a 9.5 hours ______ **b** 4.3 hours ______

c 12.6 hours ______ **d** 7.4 hours ______

e 15.7 hours ______ **f** 10.8 hours ______

g 5.25 hours ______ **h** 8.3 hours ______

5 Convert each time to hours and minutes.

a 378 minutes ______ **b** 236 minutes ______

c 173 minutes ______ **d** 685 minutes ______

e 427 minutes ______ **f** 404 minutes ______

6 What is the time:

a 6 hours 20 minutes after 10:15 p.m.? ______

b 5 hours 24 minutes before 21:30? ______

c 10 hours 4 minutes after 12:10 p.m.? ______

d 3 hours 35 minutes before midday? ______

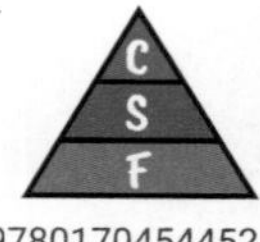

9780170454452

7 How many hours and minutes between:

a 10 a.m. and 6 p.m.? ______

b 8:35 a.m. and 5:15 p.m.? ______

c 2:40 p.m. and 8 p.m.? ______

d 5:40 a.m. and 1:20 p.m.? ______

e 15:11 and 20:24? ______

f 11:37 and 22:20? ______

WS WORKSHEET

8 What is the time:

a 233 minutes after 11:47 p.m.? ______ b 109 minutes before 2:30 p.m.? ______

c 85 minutes after 7:54 a.m.? ______ d 174 minutes before 1 p.m.? ______

e 386 minutes after 18:40? ______ f 270 minutes before 15:51? ______

9 The map below illustrates the time differences between the Australian states relative to AEST.

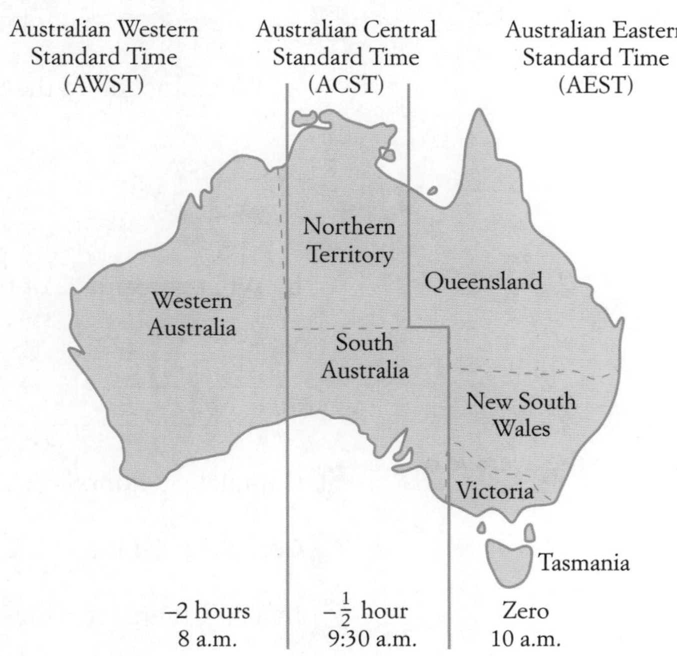

When it is 3:30 p.m. in Sydney, what is the time in:

a Hobart? ______ b Alice Springs (NT)? ______

c Perth? ______ d Brisbane? ______

e Melbourne? ______ f Adelaide? ______

10 Use the map from question **9** to help you find the time in Sydney when it is:

a 4 a.m. in Canberra ______ b 6:30 p.m. in Perth ______

c 1:15 p.m. in Darwin ______ d 10:47 a.m. in Cairns (Qld) ______

e 15:05 in Adelaide ______ f 07:20 in Hobart ______

12 RATIOS

Name:

Due date:

Parent's signature:

Part A	/ 8 marks
Part B	/ 8 marks
Part C	/ 8 marks
Part D	/ 8 marks
Total	/ 32 marks

THIS TOPIC, RATIOS, RATES AND TIME, COMBINES NUMBER WORK WITH MEASUREMENT UNITS. THIS ASSIGNMENT WILL HELP YOU PRACTISE AND REVISE IT.

HW HOMEWORK

PART A: MENTAL MATHS

Calculators are not allowed.

1 Write 5, −3, 8, −1, 0, 4 in ascending order.

2 Is 568 divisible by 6?

3 Construct a 92° angle.

4 Convert 0.6 to a percentage.

5 Write an algebraic expression for the average of a and b.

6 How many axes of symmetry does a parallelogram have?

7 Evaluate 54.2×1.1.

8 Find d.

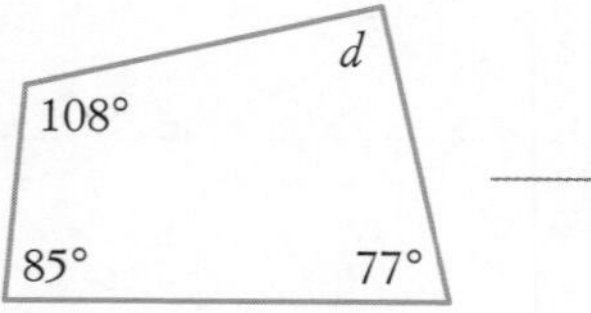

PART B: REVIEW

1 Simplify $\frac{14}{35}$. ____

2 Complete: $\frac{3}{8} = \frac{12}{\ }$.

3 In a class of 30 students, there are 12 boys.

a What fraction of the class are boys?

b What percentage of the class are girls?

4 Complete: 3 hours = ____ minutes

5 Complete: 2.4 L = ____ mL

6 In a bag, there are 5 blue, 6 red and 4 green badges. What fraction of badges:

a are red?

b are neither red nor green?

C S F

9780170454452

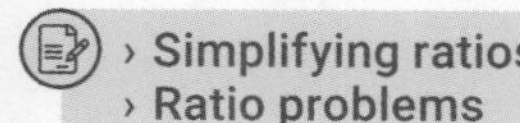

PART C: PRACTICE

› Simplifying ratios
› Ratio problems

1 For this shape, write as a simple ratio:

a shaded parts to unshaded parts

____

b unshaded parts to the whole shape.

____

2 Complete: 5 : 3 = ____________ : 18

3 Complete: 6 : 4 = 36 : ____________

4 Simplify:

a 18 : 45

____

b 20 : 8 : 16

____

c 45 cm : 1 m

____

d 60c : $2

____

PART D: NUMERACY AND LITERACY

1 In a box of chocolates, the ratio of white to dark chocolates was 2 : 3. If there were 12 white chocolates, how many dark chocolates were there?

____

2 **a** Chelsea prefers mixing cordial and water in the ratio 1 : 6. Explain what this means.

____

____

b Anton prefers to mix cordial and water in the ratio 1 : 5. Is this stronger or weaker than Chelsea's mix?

____

3 A skateboard was bought for $120 and sold for $180. Find:

a the profit made

____

b the ratio of the profit to the original price

____

c the ratio of the original price to the selling price.

____

4 To make concrete, you mix cement, gravel and sand in the ratio 1 : 2 : 4. If Libby uses 20 kg of sand, what is the mass of the total concrete mix?

____

5 The ratio of boys to girls in Tan's class is the same as the ratio of girls to boys. How is this possible?

____

____

HW HOMEWORK

C
S
F

12 RATES

I NOTICE THAT THIS ASSIGNMENT SHOWS MANY EVERYDAY APPLICATIONS OF RATES: SHOPPING, WAGES, SPEED, MAP SCALES.

Name:

Due date:

Parent's signature:

Part A	/ 8 marks
Part B	/ 8 marks
Part C	/ 8 marks
Part D	/ 8 marks
Total	/ 32 marks

HW HOMEWORK

PART A: MENTAL MATHS

Calculators are not allowed.

1 Write 7, −1, 2, −8, 3, −5 in ascending order.

2 **a** What is a reflex angle?

b Draw a reflex angle.

3 Is 296 divisible by 4?

4 Convert 45% to a simple fraction.

5 Write an algebraic expression for twice the product of a and b, decreased by c.

6 Find k.

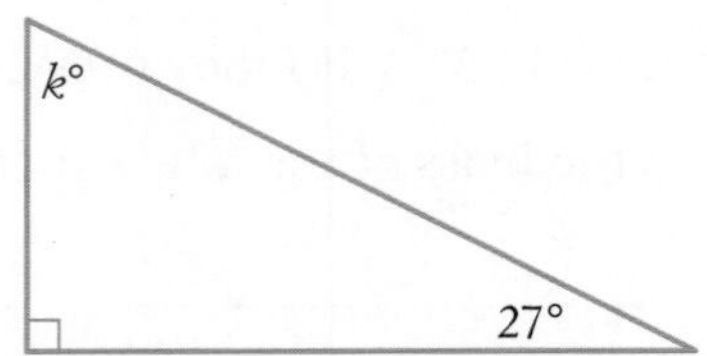

7 What fraction of the letters in the alphabet are vowels?

PART B: REVIEW

1 Emma buys 5 water bottles for $35.50.

a What is the cost of one bottle?

b What is the cost of 3 bottles?

2 Convert $15.20 into cents.

3 On a map, 2 cm represents 15 km.

a What does 1 cm represent?

b What does 5 cm represent?

4 How many times does $18.10 go into $705.90?

5 Tom earns $45.60 for 3 hours' work.

a How much does he earn for one hour's work?

b How much does he earn for 8 hours' work?

PART C: PRACTICE

› Simplifying rates
› Best buys
› Rate problems

1 Simplify each rate.

a \$3.60 for 45 minutes

b 176 m in 8 s

c 1248 students for 48 teachers

2 Which size of ice cream is the best buy?

A 2 L for \$5.28

B 1.5 L for \$3.48

C 750 mL for \$1.88

3 A parcel costs \$8.40/kg to post. How much will it cost to post a 3.2 kg parcel?

4 Khalil can type at a rate of 48 words/min.

a How many words can he type in half an hour?

b How long will he take to type a 5000-word report? Answer to the nearest minute.

5 Which brand of jam is the best value for money?

A Jim Jam: \$1.40 for 100 g

B Jam I Am: \$3.60 for 250 g

C T'rrific Jam: \$2.50 for 180 g

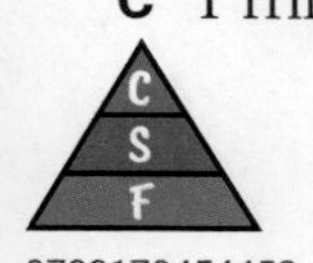

PART D: NUMERACY AND LITERACY

1 If 12 cans of cola cost \$22.20, what is the unit price?

2 Petrol costs 146.8 c/L.

a Write 'c/L' completely in words.

b How much does 53 L of petrol cost, to the nearest cent?

c How much petrol can be bought for \$60? Answer to the nearest 0.1 L.

3 Write the units used to measure each rate.

a speed

b population growth

4 Maya takes 2 hours to cycle 20 km and then walks for one hour, travelling a further 4 km. Calculate her average speed over the time.

5 How long will it take to fly 2027 km from Sydney to Alice Springs at 650 km/h? Answer to the nearest minute.

9780170454452

12 RATES AND TIME

THIS ASSIGNMENT HELPS YOU PRACTISE YOUR SKILLS WITH RATES, TRAVEL GRAPHS AND TIME. WORK HARD TO MASTER THESE IDEAS.

Name:

Due date:

Parent's signature:

Part A	/ 8 marks
Part B	/ 8 marks
Part C	/ 8 marks
Part D	/ 8 marks
Total	/ 32 marks

HW HOMEWORK

PART A: MENTAL MATHS

Calculators are not allowed.

1 Evaluate $\frac{2}{3} - \frac{1}{2}$.

2 Is 756 divisible by 9?

3 Draw a trapezium.

4 Evaluate $8 \times (-2) + (-2)$.

5 If n is an odd number, write an algebraic expression for the next odd number.

6 Write 2 angle sizes that are supplementary.

7 Evaluate 4^4.

8 Solve $\frac{k+4}{3} = 7$.

PART B: REVIEW

1 Write 8:35 p.m. in 24-hour time.

2 Write 14:20 in 12-hour time.

3 Round 8.457:

a to the nearest whole number ______

b to the nearest tenth ______

4 Complete: 4 h = ______ min.

5 Complete: 205 min= ______ h ______ min.

6 A bathroom shower pumps out water at a rate of 15 L/min.

a How much water is used in a 5-minute shower?

b How long does it take to pump out 100 L of water? Answer to the nearest second.

C S F

9780170454452

PART C: PRACTICE

› Travel graphs
› Time calculations

1 Round 4 h 48 min to the nearest hour.

____

2 Calculate the time difference between:

a 11:30 a.m. and 4:10 p.m.

____

b 09:45 and 16:25.

____

3 This travel graph shows Sam's journey.

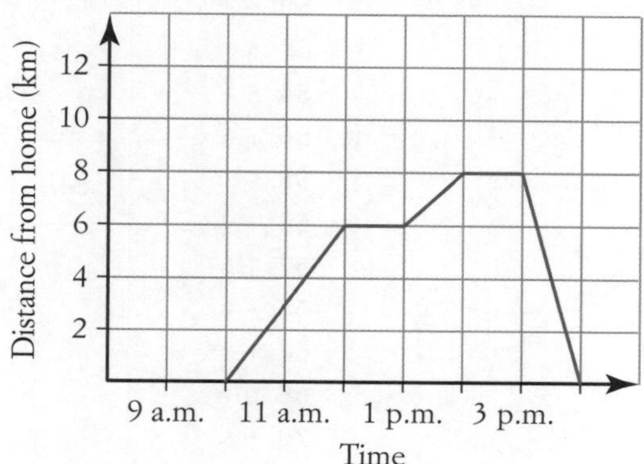

a What happened at 3 p.m.?

____

b At what time did Sam first stop?

____

c What was his speed between the 2 stops?

____

4 What is the time:

a 3 h 10 min after 7:25 p.m.?

____

b 2 h 35 min before 7:25 p.m.?

____

PART D: NUMERACY AND LITERACY

1 Name one way in which 24-hour time is better than 12-hour time.

____

____

2 This is part of a bus timetable:

Rose St	7:30	8:05	8:40
Briar Rd	7:45	8:18	8:55
Stone St	7:55	8:26	9:05
Green Rd	8:12	8:38	9:22

a How many minutes does it take the 8:05 bus from Rose St to travel to Green Rd?

____

b If I take the 8:40 bus from Rose St, where will I be in 25 minutes?

____

c If I need to be at Stone St before 8 a.m., at what time should I catch the bus from Briar Rd?

____

3 On a travel graph:

a what is shown on the vertical axis?

____

b what does it mean if the graph becomes steeper?

____

____

4 (2 marks) Draw a travel graph for this journey. Talia left home at 10 a.m. and cycled 12 km to arrive at Vanessa's home at midday. After $2\frac{1}{2}$ hours, Talia went home, arriving at 4 p.m.

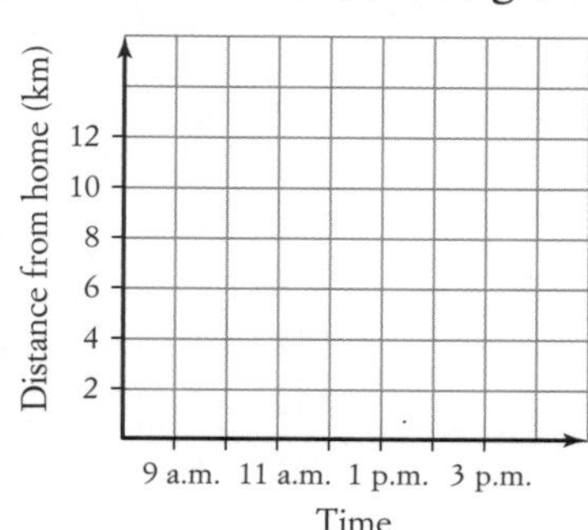

ANSWERS

Chapter 1

StartUp assignment 1 PAGE 01

1 5
2 9
3 Four hundred and nine thousand, six hundred and seventy-five
4 Teacher to check angle is obtuse (more than 90° and less than 180°).

5 **a** 6 **b** square
6 10 000
7 1, 3, 9, 27
8 5
9 10 or 11
10 90
11 105
12 $\frac{2}{3}$
13 $19
14 4:40 p.m.
15 **a** 83 **b** 16 **c** 32 **d** 49
16 5, 15, 25, 51, 52, 55
17 1 000 000
18 20
19 **a** 7 **b** 45 **c** 55 **d** 275
20 **a** true **b** false **c** false **d** true
21 770, 707, 87, 78, 77, 7
22 35
23 **a** 0 **b** 7 **c** 14 **d** 120 **e** 6
24 **a** 19 or 23 **b** 24
25 4 or −4

Challenge

Endip 94, Colin 88, Dominic 80, Benazir 76, Anh 65

Adding and subtracting integers PAGE 03

1 −2 **2** −4
3 −5 **4** −1
5 −3 **6** −1
7 −6 **8** 3
9 3 **10** −6
11 −11 **12** −5
13 −12 **14** −14
15 −8 **16** 3
17 0 **18** 1
19 −4 **20** 9
21 −1 **22** 3
23 1 **24** 0
25 3 **26** −7
27 −3 **28** 1
29 −8 **30** −5
31 −4 **32** −3
33 −14 **34** −4
35 −2 **36** −17
37 −6 **38** −18
39 −11 **40** −10
41 8 **42** 15
43 10 **44** 5
45 8 **46** 3
47 −1 **48** 2
49 −8 **50** −3
51 −3 **52** 5
53 −10 **54** 5
55 2 **56** −6
57 −1 **58** 9
59 −1 **60** −9
61 4 **62** −5
63 −11 **64** −9
65 −7 **66** −15
67 6 **68** 0
69 −4 **70** −8
71 6 **72** −12

Integer review PAGE 04

1 6, 5, 4, 0, −1, −3, −6, −8
2 **a** −2 **b** 2 **c** −2 **d** −15 **e** −8 **f** 8 **g** −8 **h** 15 **i** 2
3 **a** −2°C **b** −1°C **c** −11°C **d** 6°C **e** −3°C **f** −7°C
4 −15, −10, −7, −2, 2, 8, 11, 20
5 **a** 27 **b** −5 **c** −56 **d** −7 **e** 13 **f** 0 **g** −1 **h** 5 **i** −30 **j** −2 **k** 4 **l** −16
6 **a**

+	−2	7	0	−9	−3
1	−1	8	1	−8	−2
−5	−7	2	−5	−14	−8
8	6	15	8	−1	5
3	1	10	3	−6	0
−1	−3	6	−1	−10	−4

b

×	3	−6	2	10	−7
5	15	−30	10	50	−35
−4	−12	24	−8	−40	28
−2	−6	12	−4	−20	14
7	21	−42	14	70	−49
−1	−3	6	−2	−10	7

7 **a** 12°C **b** 8°C **c** −7°C **d** −4°C **e** 5°C **f** 8°C

8 **a** −42 **b** −12 **c** −2 **d** −64 **e** 5 **f** 25 **g** −60 **h** 23 **i** −1 **j** 2 **k** −7 **l** −31

Chapter 2

StartUp assignment 2 PAGE 12

1 **a** 390 **b** 32

2 4

3 No

4 1620

5 century

6 128

7

8 1, 2, 3, 4, 6, 9, 12, 18, 36

9 8.2

10 8

11 12 cm^2

12 13 900 528

13 1000

14 8:40 p.m.

15 south-west

16 **a** 180 **b** 270 **c** 360

17

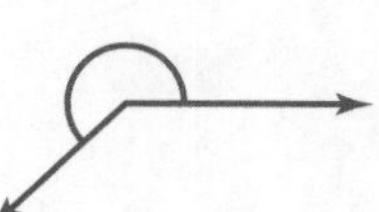

18 **a** 108° **b** 45° **c** 52° **d** 120°

19 124°

20 right angle

21 acute

22

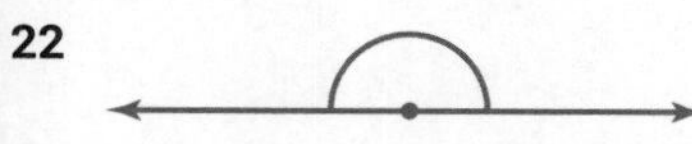

23 *C*

24 **a** parallelogram

b Sides that are parallel are marked with the same numbers of arrows.

25 8

26 **a** equilateral **b** 60°

27

49°

28 reflex

29 True

30 *F* and *H*

31 **a** square **b** 90°

Challenge 170

Find the unknown angle PAGE 16

1 $a = 142, b = 38$

2 $x = 48$

3 $y = 202$

4 $a = 100$

5 $c = 115$

6 $x = 72$

7 $a = 70$

8 $p = 20$

9 $a = 50, b = 140$

10 $x = 22$

11 $a = c = 114, b = d = 66$

12 $h = 50, k = 75$

13 $x = 45$

14 $k = 40$

15 $r = 30, s = 60, t = 150$

16 $m = 55, n = 55$

17 $x = 95, y = 85, z = 95$

18 $z = 65, x = 65, y = 50$

19 $p = 120$

20 $x = 130$

Angles crossword PAGE 18

Across

3 protractor

8 complementary

10 transversal

12 vertex

13 four

16 parallel

17 angles

18 degree

20 turn

22 straight

24 reflex

25 two

26 scale

27 ninety

Down

1 arm

2 supplementary

4 revolution

5 co-interior

6 perpendicular

7 intersection

8 corresponding

9 alternate

11 vertically

14 obtuse

15 adjacent

17 acute

19 equal

21 sixty

Chapter 3

StartUp assignment 3 PAGE 30

1 500

2 **a** 70 **b** 314.2 **c** \$3.15

3 $\frac{3}{10}$

4 736 000

5 360

6 16

7 70 000
8 28 cm
9 1815
10 26
11 0.75
12 31
13 isosceles
14 12 746
15 5216, 7352, 9004, 10 840
16 **a** 23 **b** 125 **c** 9 **d** 153
17 11, 13, 17, 19
18 even
19 55
20 $8 + 10 - (3 \times 3 + 6) = 3$
21 **a** true **b** false **c** false **d** true
22 **a** four hundred and sixty one thousand, and twenty-nine
b 0
23 teacher to check
24 5
25 4 or −4
26 49
27 **a** 72 **b** 22 **c** 20 **d** 150

Challenge:

1	15	14	4
12	6	7	9
8	10	11	5
13	3	2	16

Powers and roots

PAGE 32

1 **a** 5^4 **b** 2^7
c 7^3 **d** 16^2
e 3^8 **f** $(2.5)^5$
g $(-9)^3$ **h** $\left(\frac{2}{3}\right)^6$
2 **a** $3 \times 3 \times 3 \times 3$ **b** 5×5
c $7 \times 7 \times 7 \times 7 \times 7$ **d** $11.2 \times 11.2 \times 11.2 \times 11.2$
e $(-8) \times (-8) \times (-8) \times (-8) \times (-8) \times (-8) \times (-8)$
f $1\frac{1}{2} \times 1\frac{1}{2} \times 1\frac{1}{2}$
3 **a** 169 **b** 729
c 64 **d** 125
e 16 **f** 100
g 128 **h** 1 000 000
i 1 **j** 1.44
k 194 481 **l** −2187
m 441 **n** −3125
o 12.167 **p** 59 049
4 **a** 7 **b** 13
c 31 **d** 2.8
e 1 **f** 180
g 9 **h** 2
i −2 **j** 1.7
k −1.7 **l** −0.6
5 **a** 6.32 **b** 8.66
c 4.48 **d** −2.29
6 **a** yes **b** yes
c no **d** yes
e yes **f** yes
7 **a** 4 **b** 8
c 16 **d** 32
e 64 **f** 128
g 256 **h** 512
8 **a** 100 **b** 1 000 000
c 1 000 000 000 **d** 10 000 000
e 1000 **f** 10 000
g 10 000 000 000 **h** 100 000
9 100 000 000 000 000
10 **a** 18 **b** 256
c 7 **d** 1331
11 **a** 5 **b** 6
c 5 **d** 4

Factor trees

PAGE 34

1
22
2 × 11
$22 = 2 \times 11$

2
38
19 × 2
$38 = 19 \times 2$

3
65
5 × 13
$65 = 5 \times 13$

4
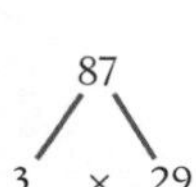
$87 = 3 \times 29$

5
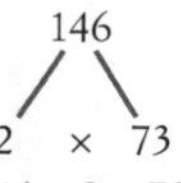
$146 = 2 \times 73$

6
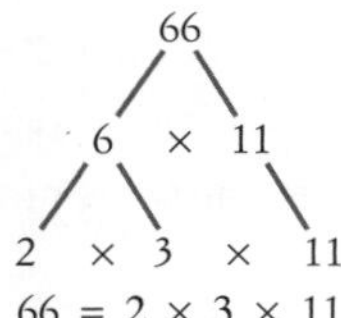
$66 = 2 \times 3 \times 11$

7

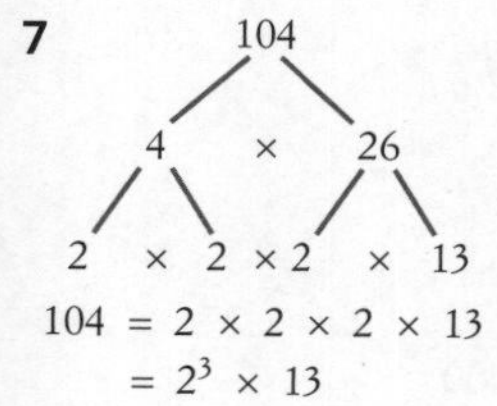

$104 = 2 \times 2 \times 2 \times 13$
$= 2^3 \times 13$

8

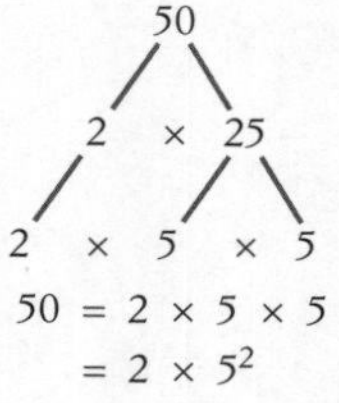

$50 = 2 \times 5 \times 5$
$= 2 \times 5^2$

9

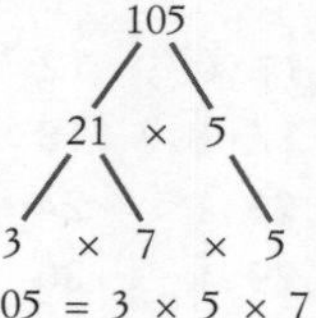

$105 = 3 \times 5 \times 7$

10

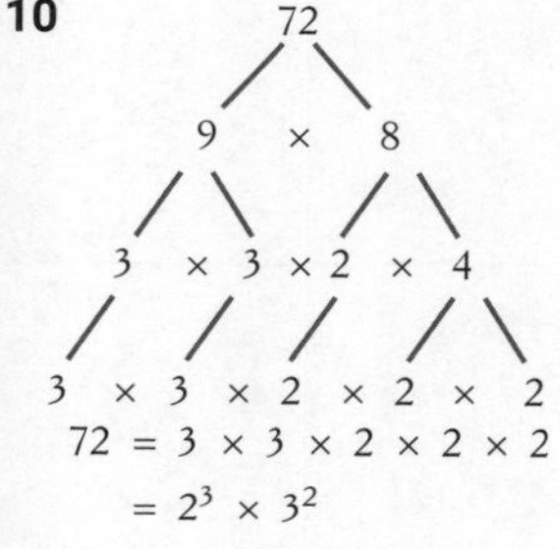

$72 = 3 \times 3 \times 2 \times 2 \times 2$
$= 2^3 \times 3^2$

11

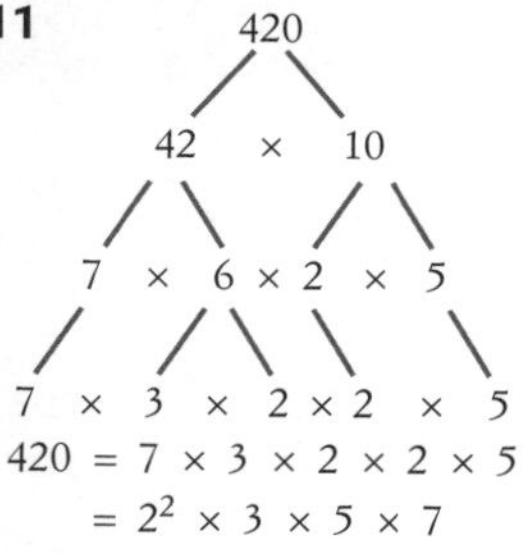

$420 = 7 \times 3 \times 2 \times 2 \times 5$
$= 2^2 \times 3 \times 5 \times 7$

12

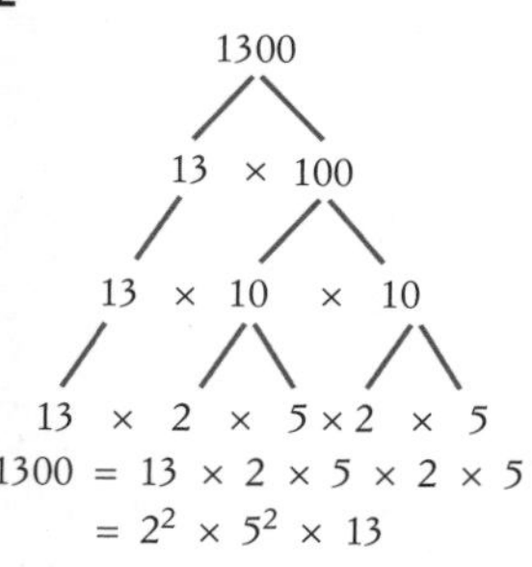

$1300 = 13 \times 2 \times 5 \times 2 \times 5$
$= 2^2 \times 5^2 \times 13$

13

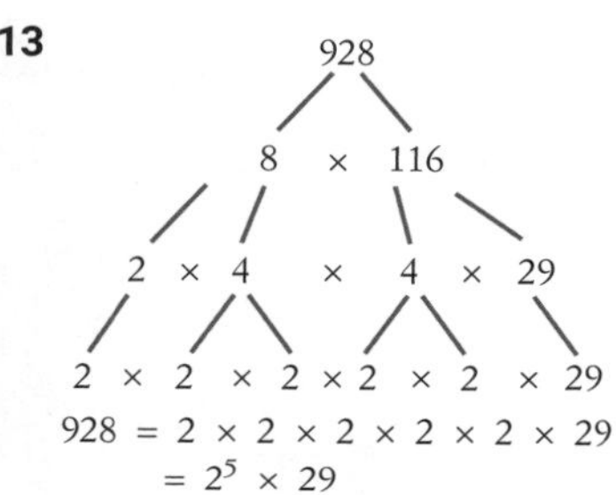

$928 = 2 \times 2 \times 2 \times 2 \times 2 \times 29$
$= 2^5 \times 29$

Cross number puzzle

PAGE 44

Across

1 36
2 1001
4 21
6 64
7 27
9 342
11 32
13 84
14 11155
16 5235
19 81
20 444
22 198
25 11
27 49
28 12
29 9876
31 18
32 1219
33 30

Down

1 366
2 144
3 12345
5 12345
6 63
8 72
10 235
12 1124
13 831
15 54
17 288
18 114
20 4091
21 411
23 9999
24 1760
26 121
30 73

Chapter 4

StartUp assignment 4 PAGE 45

1 **a** 3500 **b** 253
c 20 **d** 96
2 24 cm^2
3 **a** 0.25 **b** 25%
4 $y = 65$
5 240 200
6 **a** −4 **b** 34
c 256
7 9^4
8 0.4
9 parallelogram
10 **a** 9 **b** 32
c 30 **d** 7
11 8, 16, 24, 32, 40
12 $\frac{7}{16}$
13 **a** $\frac{67}{100}$ **b** $\frac{1}{5}$
14 **a** 1, 2, 4, 8, 16 **b** 1, 3, 5, 9, 15, 45
15 No
16 **a** 4 **b** 5
17 6
18 $\frac{11}{27}$
19 **a** 240 **b** 2000
c 2400 **d** 72
20 **a** 5 **b** 7
c 16 **d** 6
21 Yes
22 40
Challenge: 7

Equivalent fractions PAGE 48

1 H	**2** T	**3** N	**4** X	**5** E	**6** G
7 L	**8** S	**9** W	**10** Z	**11** C	**12** A
13 J	**14** Q	**15** V	**16** E	**17** B	**18** A
19 T	**20** K	**21** Y	**22** D	**23** P	**24** I
25 R	**26** O	**27** M	**28** U	**29** R	**30** F

What is the name of the line that separates the numerator from the denominator of a fraction?

Answer: Vinculum.

Percentages cross number PAGE 47

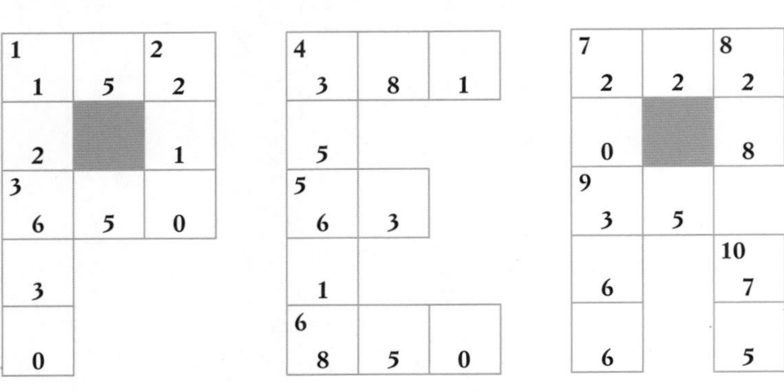

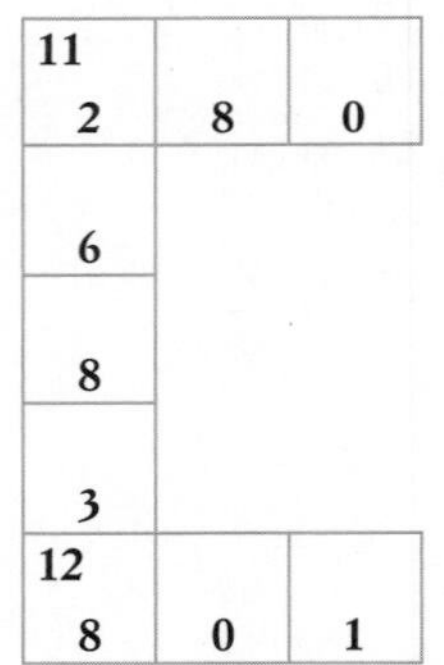

9780170454452

Fractions and percentages crossword PAGE 58

Across

1 simplify
4 vinculum
5 proper
9 equivalent
10 fifty
12 improper
14 mixed
16 twenty
17 denominator

Down

2 percent
3 quarter
6 hundred
7 numerator
8 half
11 third
13 reciprocal
15 fraction
18 reduce

Chapter 5

StartUp assignment 5 PAGE 59

1 1500
2
3 7610
4 5.8
5 true
6 20°
7 160
8 obtuse
9 4 thousands
10 \$24
11 0.02
12 3600
13 134 cm
14 $\frac{3}{5}$
15 22 cm
16 a −5 **b** 14
c −13 **d** 3
17 a 5 **b** −1
c 12
18 a 10 **b** 9
c 3 **d** 8
19 a 43 **b** 36
c −7
20 a 24 **b** 5
c 11 **d** 7
21 a 10, 13, 3, 4, 4, 16 **b** increases by 3
22 true
23 a 27 **b** 5
c 18 **d** −4

Challenge: A = 9, M = 8, B = 2, C = 1

Substitution PAGE 61

1 a 7 **b** −1
c 16 **d** 2
e 14 **f** 6
g 13 **h** −6
2 a 2 **b** −12
c −11 **d** −18
e 8 **f** 3
g 10 **h** −13
3 a 0 **b** −4
c 27 **d** 10
e −4 **f** −20
g 12 **h** −6
4 a 13 **b** −17
c 3 **d** 20
e 48 **f** 6
g 12 **h** 6
5 a 2 **b** −16
c 11 **d** 190
e 1 **f** 30
g −100 **h** −59
i 1 **j** 31
k 13 **l** −63
m −149 **n** 105
o −8 **p** 27
q 250 **r** 220
s 27 **t** $\frac{13}{20}$

Equations match PAGE 62

1 L **2** M **3** A **4** B **5** E **6** D
7 F **8** N **9** J **10** H **11** G **12** O
13 C **14** I **15** K

Chapter 6

StartUp assignment 6 PAGE 72

1 2110
2 a −14 **b** −6
3 800
4 a A number with 2 factors, 1 and itself.
b 11, 13, 17 or 19
5 a Triangular prism **b** 5
6 $150 = 2 \times 3 \times 5^2$
7 a 0.05 **b** $\frac{1}{20}$
8 a 3600 **b** 169
9 30
10 straight angle
11 a One solution to this problem is shown below:

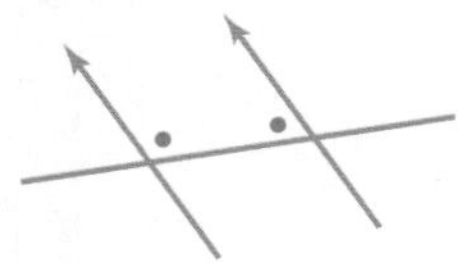

b They are supplementary (sum to 180°).

12

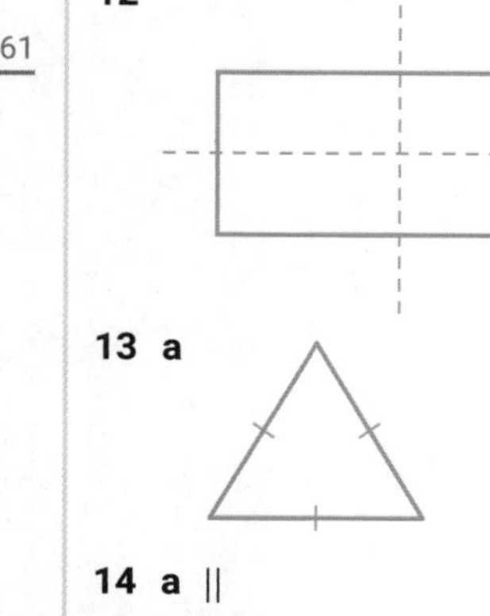

13 a

b yes

14 a $\parallel$ **b** $\perp$
15 a 63° **b** Acute **c** $\angle ZOH$ or $\angle HOZ$

16 6

17 spin

18 40°

19 **a** regular octagon **b** pentagon

20 **a**

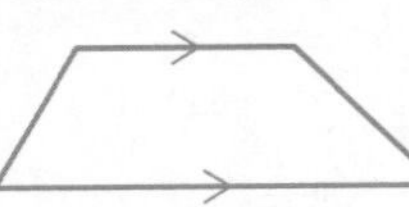

b

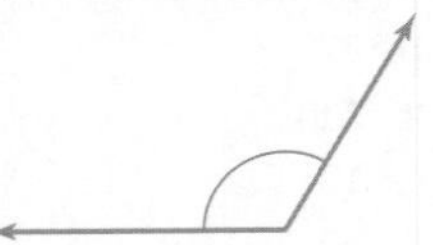

21 $y = 120°$

22 **a** A four-sided polygon

b Square, rectangle, rhombus, trapezium, parallelogram or kite

23 **a** 1 **b** 0

24 **a** An angle that is between 90° and 180°

b

Challenge: 204

Transformations

PAGE 74

1 **a** reflection **b** translation

c reflection **d** rotation

e translation **f** rotation

g translation **h** rotation

i reflection **j** rotation

k reflection **l** rotation

2 **a**

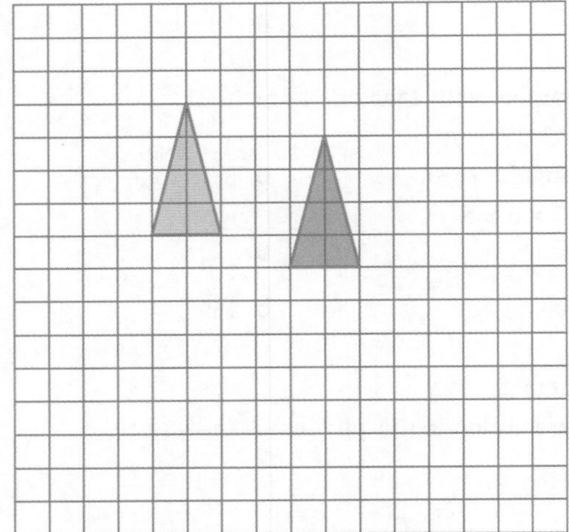

b

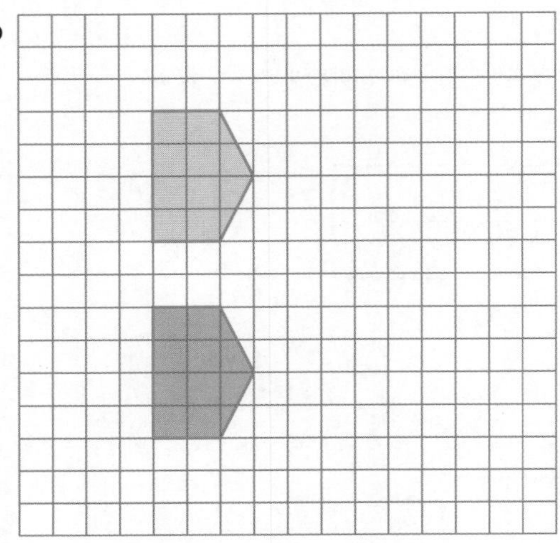

c

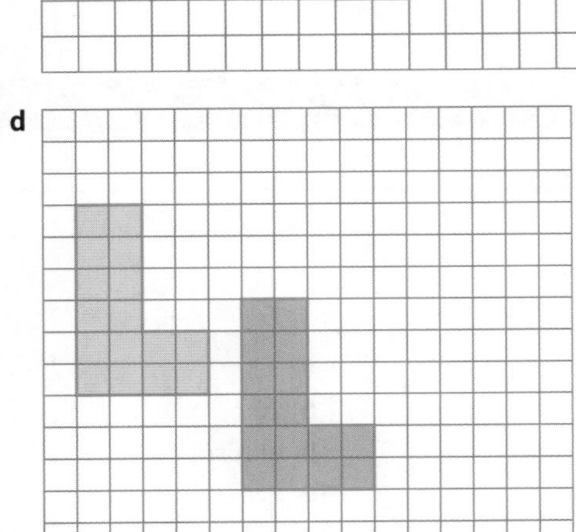

d

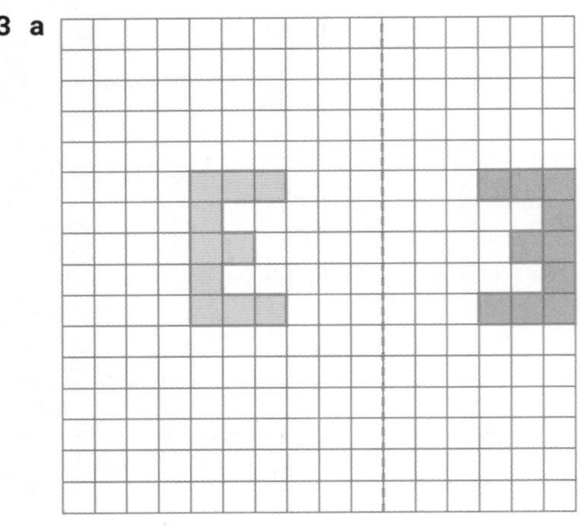

3 **a**

b

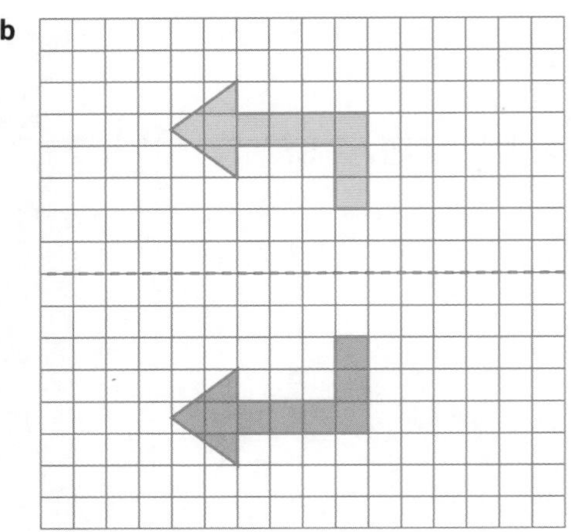

9780170454452

c

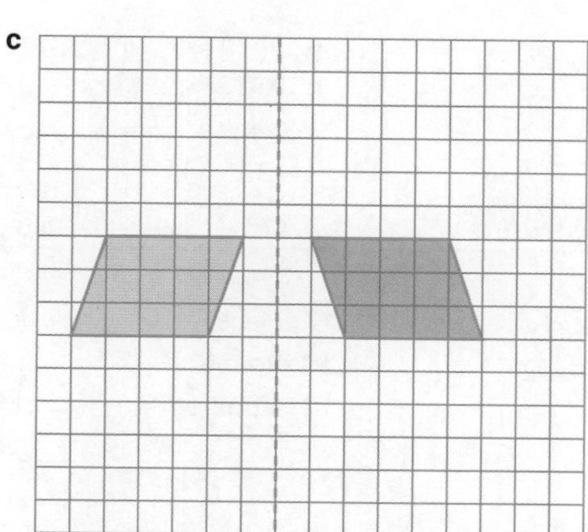

d

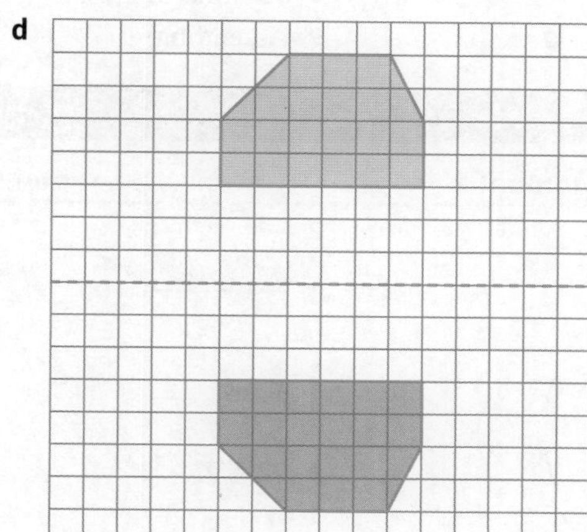

4 a

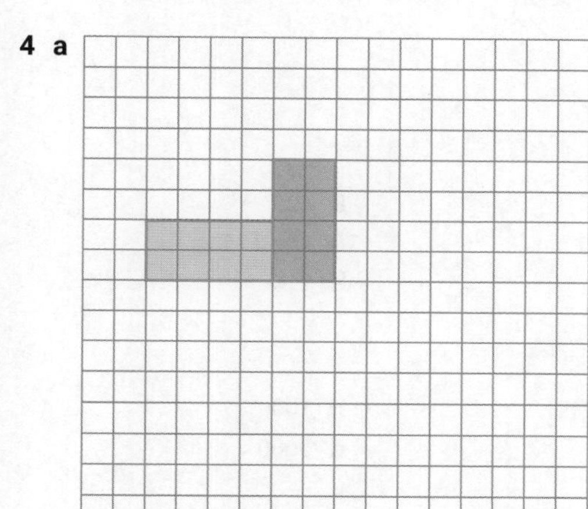

b

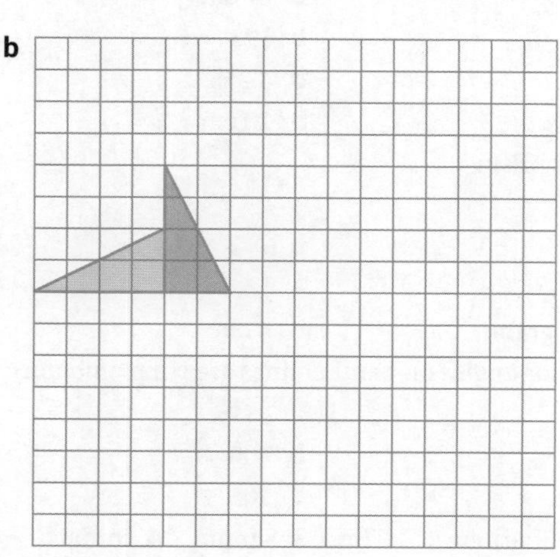

c

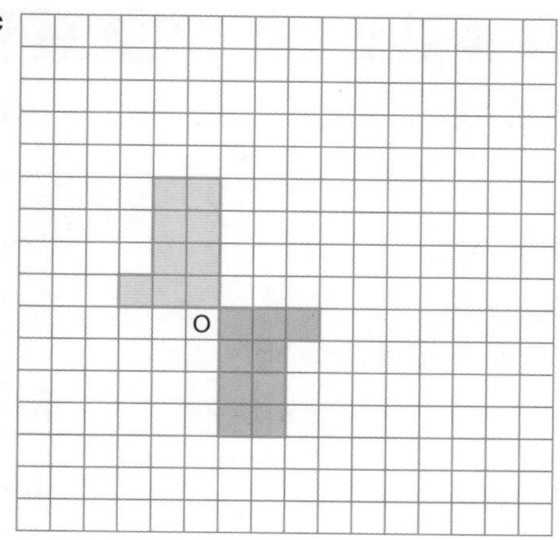

d

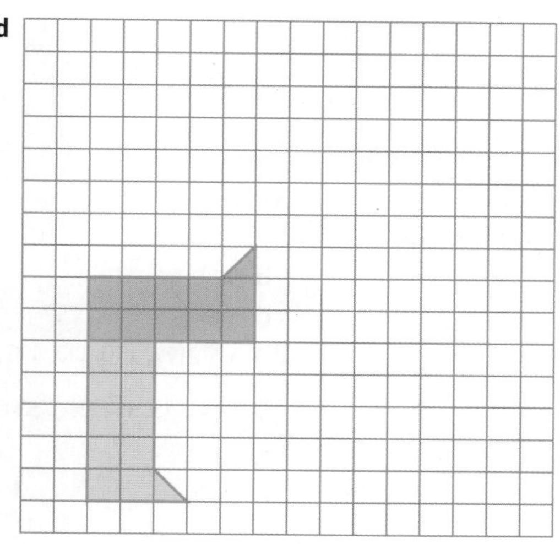

Symmetry

PAGE 77

Axes of symmetry, order of rotational symmetry:

1 2, 2	**2** 6, 6	**3** 5, 5	**4** 6, 6
5 8, 8	**6** 1, 0	**7** 2, 2	**8** 4, 4
9 3, 3	**10** 3, 3	**11** 4, 4	**12** 2, 2
13 0, 2	**14** 2, 2	**15** 3, 3	**16** 0, 2
17 4, 4	**18** 5, 5		

Triangle geometry

PAGE 78

1 $r = 130$	**2** $a = 120$	**3** $k = 45$
4 $x = 60, y = 3$	**5** $q = 36$	**6** $m = 55$
7 $h = 85$	**8** $a = 70, b = 110$	**9** $e = 136$
10 $v = 33$	**11** $m = 10$	**12** $p = 20$
13 $y = 120$	**14** $c = 30$	**15** $x = 50, y = 80$
16 $c = 50, d = 75$	**17** $z = 120$	**18** $t = 80, u = 100$
19 $p = 110$	**20** $y = 60 + 55 = 115$	**21** $n = 35$
22 $k = 35, i = 5$	**23** $h = 65$	**24** $y = 67.5$

Chapter 7

StartUp assignment 7 PAGE 91

1 a −6 **b** 400
c 19
2 1, 11
3 a
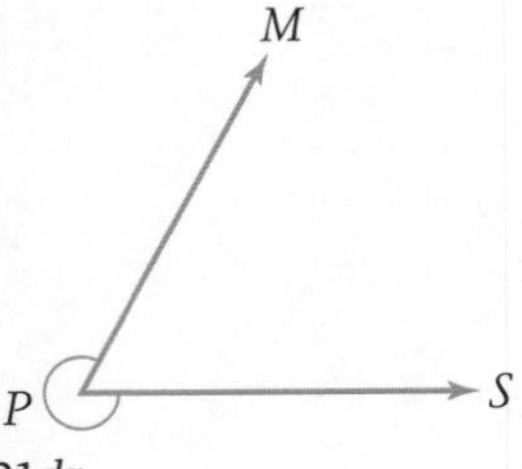
b 0
4 Co-interior angles are supplementary.
5 $4 \times 4 \times 4$ **6** $4\frac{3}{4}$
7 40
8
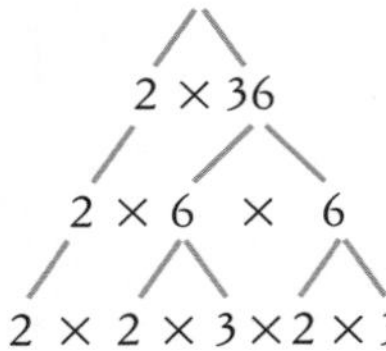

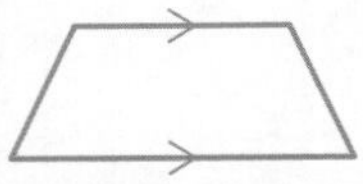

9 $21dr$ **10** 60°
11 Its last digit is 5 or 0.
12 a 560 **b** 96
c 1019 **d** 143
13 72, 74, 76 **14** 177, 171, 170, 117, 111, 107
15 $\frac{7}{10}$ **16** 344, 345, 354, 355, 534, 543
17 a 1900 **b** 84
c 852 **d** 3510
18 a 2 **b** 1 hundred
19 a 75 200 **b** 75 000
20 $\frac{9}{20}$
21 a 813 **b** 829
c 137 **d** 480
e 18 000
22 a $<$ **b** $>$
c $<$
Challenge: 10 × \$1, 15 × 20c, 25 × 10c

Where's the point? PAGE 93

1 a 4560 **b** 96 330
c 41 400 **d** 3472 000
e 3042 000 **f** 131.1
g 7334.6 **h** 4.56
i 4.14 **j** 144
k 733.46 **l** 1.311
m 1311 **n** 3.472
o 4340 **p** 43.4
q 2300 **r** 230
s 4.34 **t** 43.4
3 a 117.6 **b** 412.65
c 0.8 **d** 120
e 15 750 **f** 7840
g 157 500 **h** 1.176
i 0.8 **j** 412.65
k 1176 **l** 209 600
m 2096 **n** 11.76
o 1.575 **p** 0.315
q 500 **r** 2.4
s 490 **t** 13.1

Decimals crossword PAGE 104

Across
5 Descending **7** Place
8 Thousandth **10** Round
11 Recurring **12** Hundredth
13 Estimate

Down
1 Fraction **2** Point
3 Decimal **4** Terminating
6 Tenth **9** Ascending

Chapter 8

StartUp assignment 8 PAGE 105

1 $3mp^2$
2 (diagram of an obtuse angle)
3 7, 14, 21, 28, 35 **4** $x = 6$
5 31
6 a

72
2 × 36
2 × 6 × 6
2 × 2 × 3 × 2 × 3

b $72 = 2^3 \times 3^2$
7 2, 3 **8** 625
9 $<$ **10** $x = 115°$
11 a −12 **b** 216
c $\frac{11}{40}$
12 −13
13 a 1000 **b** 1000
c 10 **d** 1000
14 5 mm
15 a 20 cm **b** 21 cm^2
16 a 385.2 **b** 195
c 20.74 **d** 4.356
17 a 5 cm **b** 20 cm
18 rectangular prism
19 24 cm^2
20 a 1 **b** 7
c 47
21 a parallelogram **b** 4 cm
c Co-interior angles on parallel lines are supplementary.
d 105
22 a 1700 **b** 400
c 210 **d** 105
Challenge: 1 Charlotte **2** Tim **3** Sammi **4** Jordan

Areas on a grid
PAGE 107

1 8 cm^2	**2** 11 cm^2
3 9 cm^2	**4** 20 cm^2
5 17 cm^2	**6** 8 cm^2
7 4 cm^2	**8** 12 cm^2
9 6 cm^2	**10** 9.5 cm^2
11 12 cm^2	**12** 14 cm^2
13 10 cm^2	**14** 7.5 cm^2
15 6 cm^2	**16** 13.5 cm^2
17 6 cm^2	**18** 8 cm^2

Composite areas
PAGE 108

1 30 cm^2	**2** 36 cm^2
3 24 cm^2	**4** 78 cm^2
5 30 cm^2	**6** 13 cm^2
7 40 cm^2	**8** 36 cm^2
9 27.5 cm^2	**10** 28 cm^2
11 84 cm^2	**12** 50 cm^2
13 29.5 cm^2	**14** 1600 cm^2
15 34 m^2	**16** 2575 mm^2
17 19.85 m^2	**18** 3 070 000 mm^2

Volume and capacity
PAGE 110

1 750 cm^3	**2** 36 cm^3
3 **a** 56 cm^2	**b** 24 cm^3
4 80 cm	**5** 5 cm
6 **a** 28.875 cm^3	**b** 346.5 cm^3
c Many answers possible	**d** In one long pile
7 $16\frac{2}{3}$ cm	**8** 7.8 g
9 643.5 L	**10** 125 000
11 7	

Challenges: 1 21 cm^3 **2** 7 cm × 8 cm × 9 cm

Area and volume crossword
PAGE 112

Across

2 breadth	**3** square
6 area	**8** micro
14 height	**16** parallelogram
19 kilolitre	**21** centi
22 milli	**23** length
24 prism	

Down

1 perpendicular	**4** rectangle
5 triangle	**7** composite
8 mega	**9** perimeter
10 capacity	**11** metric
12 volume	**13** cross
15 tonne	**17** metre
18 litre	**19** kilo
20 hectare	**21** cubic

Chapter 9

StartUp assignment 9
PAGE 120

1 $d = 155, e = 25$

2 17.85

3 7:03 p.m.

4 **a** 360 **b** $\frac{2}{3}$ **c** 81 **d** \$65

5 2025 (or 2034 if 2025 has passed)

6 d

7 Angles that add up to 90°

8 $x = 107$

9 $p + 4$

10 $\frac{9}{20}$

11 A quadrilateral with equal sides and equal angles (90°)

12 **a** secondary school **b** station or car park **c** golf course

13 **a** D3 **b** D1 or E1 **c** F2

14 −3, 0, 2, 3

15

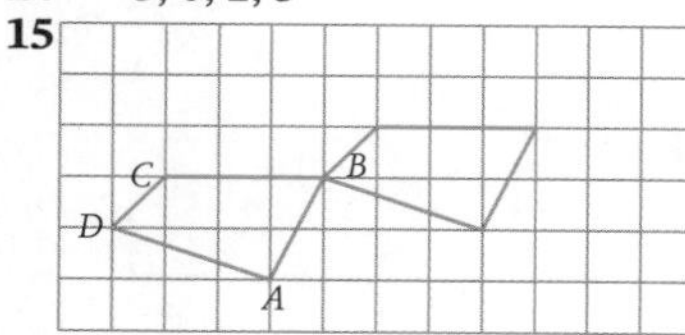

16

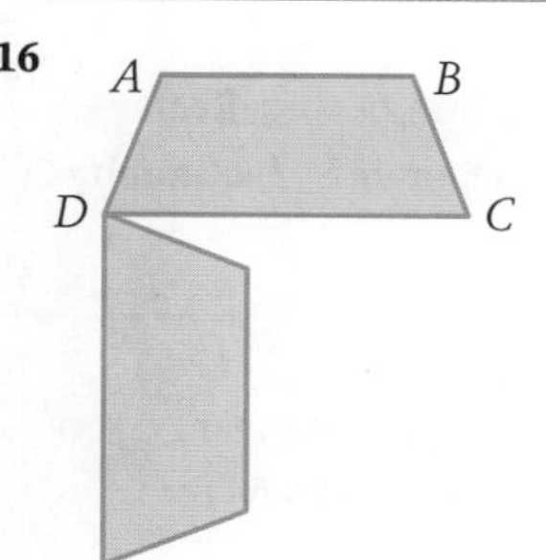

17

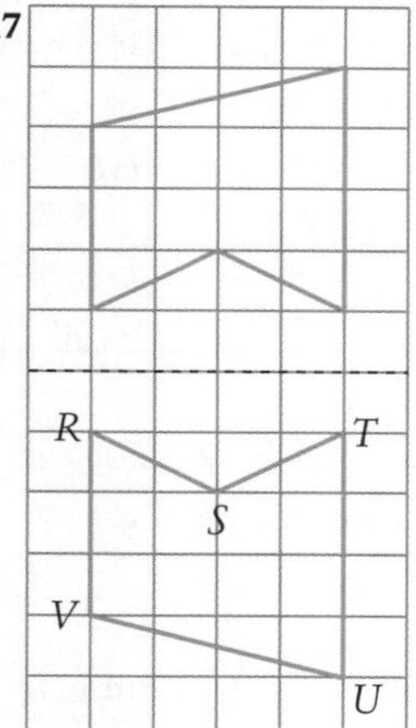

18 5

19 $A(-2, 1), B(0, 3), C(3, 1), D(2, -1)$

20 B

21

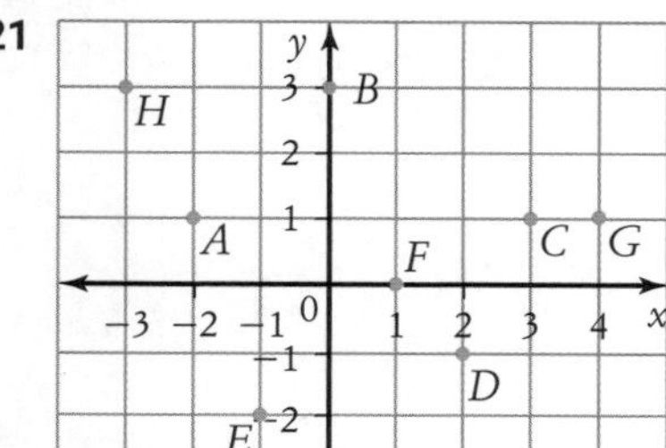

22 translation, rotation

23 (−3, −3), (2, 7) **24** 1 **25** (0, 0)

Challenge Steve is 40 and Linda is 30.

Big top employee PAGE 122

1

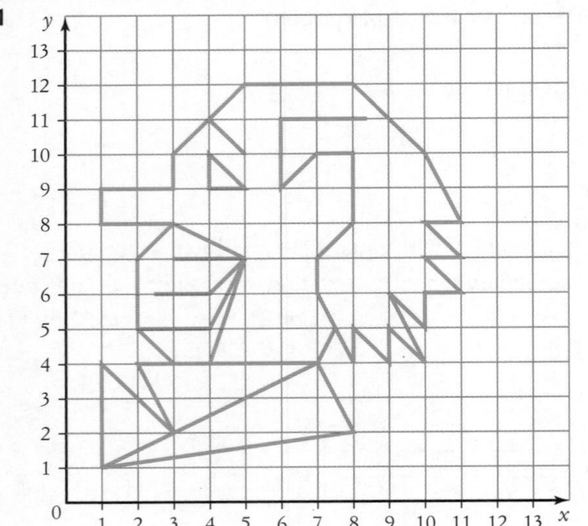

2 **a** Square **b** Trapezium **c** Rectangle
d Right-angled triangle **e** Parallelogram **f** Rhombus

The number plane PAGE 124

1 **a** 2 **b** x-axis
c vertical, up, y-axis **d** position, number
e coordinates **f** x-coordinate, 5
g cross, origin

2

$C(4, -4)$	$H(-4, 0)$	$N(0, 5)$	$A(-3, -3)$	$T(-5, 4)$
$G(-2, 3)$	$Q(-1, -4)$	$D(0, 1)$	$K(-3, -6)$	$Y(-1, 3)$
$M(0, -5)$	$J(-3, 5)$	$B(3, -1)$	$P(-2, 0)$	$S(-6, -2)$
$U(-3, -5)$	$E(3, 4)$	$I(0, -1)$	$V(5, 6)$	$L(6, -2)$
$F(4, 2)$	$R(2, -6)$	$O(0, 0)$	$X(-5, 1)$	

Number plane crossword PAGE 125

Across

1 origin **3** first **6** axes **7** Cartesian
8 axis **10** rotation **12** number **13** third
14 plane **15** line

Down

2 reflection **4** translation **5** quadrant **9** coordinate
11 values **14** pair

Chapter 10

StartUp assignment 10 PAGE 131

1 **a** alternate angles **b** 110
2 $\frac{2}{3}$
3 360°
4 **a** 6 **b** $\frac{11}{12}$
c 216 **d** \$90
5 $2k - 10$
6 $21de$
7 **a** 30 m^2 **b** 26 m
8 **a** 1200 **b** 1.2
9 (7, 3)
10 **a** sector graph **b** divided bar graph
c line graph **d** column graph
11 **a** 9 **b** 6, 7, 10, 10, 11, 13, 13, 13, 16
c 13 **d** 11
e 10 **f** 11
12 **a** 82, 84, 86, 88, 92 **b** 16, 20, 28, 32
13 $\frac{1}{3}$
14 **a** Guinea pig **b** 200
c Fish **d** Cats, dogs
e

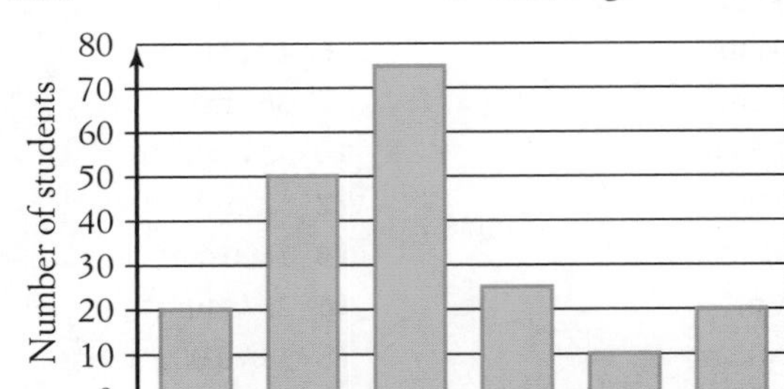

15 17
16 **a** $\frac{3}{4}$ **b** 75%
17 $y = 45$
18 **a** 2 **b** $\frac{1}{9}$
c Number of TV sets owned at home

Number of TVs
1
2
3
4
Over 4
= 2 students

Challenge:

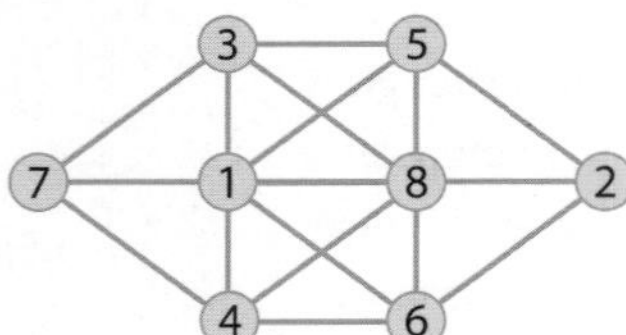

Other solutions possible

9780170454452

Mean, median, mode

PAGE 133

Q	Mean	Median	Mode(s)	Range
1	14.71	15	11	13
2	18.88	18	10 & 16	18
3	11.18	11	14	17
4	4.67	4	1	12
5	25.4	22.5	20	22
6	31.1	30	40	48
7	62.89	64	64	16
8	3.14	3	2 & 5	4
9	87	87	no mode	8
10	30	30	no mode	40
11	4.46	4	3 & 4	7
12	17	17	17	6
13	7	7	7	6
14	30.86	20	17	83
15	9.18	9	8	5
16	22.38	22.5	21 & 23	5
17	49.57	50	48 & 51	5
18	9.17	9	9	1
19	2	2	1	5
20	5.62	6	6	7

Stem-and-leaf plots

PAGE 134

1 **a** 166 cm **b** 166 cm
c 28 cm **d** 166.68 cm

2

Girls	Stem	Boys
8	15	4 5 7 8
8 6 5 3 1	16	1 4 5 5 6 6 6 7 8
5 3 2 0 0	17	0
1	18	1 2

3 **a** Girls **b** Boys

5

	Girls	Boys
Mode	170	166
Median	169	165.5
Mean	168.5	165.3
Range	23	28

6 **a** 6 **b** 3

7 Girls taller on average; boys greater spread

8 Teacher to check

Data crossword

PAGE 144

Across

4 statistics **6** cluster
8 middle **10** leaf
11 spread **12** mean
13 dot plot

Down

1 outlier **2** stem
3 median **5** average
6 column **7** range
9 data **11** sector
12 mode

Chapter 11

StartUp assignment 11

PAGE 145

1 **a** 216 **b** 11
c $\frac{1}{10}$

2 $x = 130$

3 $y = 10$

4 0.6131, 0.613, 0.6103, 0.6013

5 **a** -2 **b** 21

6 4

7 180°

8 27 cm^3

9 18

10 9 cm

11 **a** *a* and *b* (other answers possible)
b *a* and *c* (or *b* and *d*)

12 **a** $\frac{3}{4}$ **b** $\frac{9}{16}$
c $\frac{5}{6}$ **d** 15

13 **a** 2 **b** 6

14 Win, lose or draw

15 **a** 87.5% **b** 5%

16 Any impossible event (teacher to check)

17 **a** $\frac{3}{4}$ **b** $\frac{3}{10}$
c 0.835

18 $\frac{1}{2}$ or 50%

19 32%

20 a $\frac{7}{20}$ **b** $\frac{22}{25}$

21 Teacher to check

22 red, amber/yellow, green

23 a 8 **b** green

Challenge: 5

Games of chance

PAGE 147

1 a $\frac{1}{52}$ **b** $\frac{1}{2}$

c $\frac{1}{4}$ **d** $\frac{1}{13}$

e $\frac{2}{13}$ **f** $\frac{3}{13}$

g $\frac{3}{4}$ **h** $\frac{5}{13}$

i $\frac{7}{13}$

2 a $\frac{1}{24}$ **b** $\frac{1}{24}$

c $\frac{1}{12}$ **d** $\frac{11}{12}$

3 a $\frac{1}{4}$ **b** $\frac{3}{36}$

c $\frac{1}{6}$ **d** $\frac{1}{24}$

4 a $\frac{13}{51}$ **b** $\frac{4}{17}$

c $\frac{1}{17}$ **d** $\frac{1}{51}$

5 a $\frac{1}{6}$ **b** 0

c $\frac{1}{2}$ **d** $\frac{1}{2}$

e $\frac{5}{6}$ **f** $\frac{1}{3}$

6 a $\frac{1}{7}$ **b** $\frac{2}{7}$

c $\frac{1}{7}$ **d** $\frac{2}{7}$

e $\frac{4}{7}$ **f** $\frac{5}{7}$

7 36°

8 CASH in any six sectors

9 Teacher to check

10 a $\frac{1}{37}$ **b** $\frac{18}{37}$

c $\frac{18}{37}$ **d** $\frac{18}{37}$

11 a $\frac{12}{37}$ **b** $\frac{3}{37}$

c $\frac{12}{37}$ **d** $\frac{6}{37}$

e $\frac{4}{37}$ **f** $\frac{24}{37}$

12 $\frac{10}{37}$

13 $\frac{14}{37}$

14 $\frac{28}{37}$

Coins probability

PAGE 150

1

1st coin	2nd coin	No. of heads
H	H	2
H	T	1
T	H	1
T	T	0

2

No. of heads	Probability	Probability (%)
0	$\frac{1}{4}$	25
1	$\frac{1}{2}$	50
2	$\frac{1}{4}$	1

3 Each event is not equally likely, there are 2 ways of getting 1 head.

7

1st coin	2nd coin	3rd coin	No. of heads
H	H	H	3
H	H	T	2
H	T	H	2
H	T	T	1
T	T	T	0
T	H	T	1
T	T	H	1
T	H	H	2

8

No. of heads	Probability	Probability (%)
0	$\frac{1}{8}$	12.5
1	$\frac{3}{8}$	37.5
2	$\frac{3}{8}$	37.5
3	$\frac{1}{8}$	12.5

9 a 1 and 2 **b** 0 and 3

15 a 16 **b** 2

Probability crossword

PAGE 156

Across

2 decimal **4** chance
6 improbable **11** dice
13 impossible **16** probability
17 simulation **18** win
19 even **20** random
22 fraction **25** sum
28 equally **29** suit (or coin)
30 probable **33** event
34 spades **35** sample
36 fifty **37** spinner
38 space

Down

1 Buckleys **3** likelihood
5 clubs **7** experiment
8 die **9** percentage
10 frequency **12** complementary
14 diamonds **15** loss
20 range **21** possible
23 hearts **24** certain
26 unlikely **27** likely
31 heads **32** coin (or suit)

Chapter 12

StartUp assignment 12

PAGE 157

1 a 32.5% **b** $\frac{13}{40}$
2 x-axis
3 $\frac{4n}{3}$ **4** 752 or 756
5 a 2.68 **b** 16.2
c 160 **d** −4
6 a scalene **b** right-angled
7 6 cm^2 **8** trapezium
9 9 **10** $x = 60$
11 $\frac{1}{50}$
12 a Twenty minutes to ten **b** 9:40
13 1650 h **14** 9:25 p.m.

15 a $\frac{1}{5}$ **b** $\frac{3}{5}$
16 a 180 **b** 120
c 2400 **d** 850
17 a 6 p.m. **b** 8 a.m.
18 Teacher to check (124 in 2020)
19 28
20 a 25 **b** 18
21 a 8 **b** 6
22 8 **23** $\frac{6}{17}$
24 a 48 **b** $266
25 7 h 30 min **26** $1.45

Challenge: Run both hourglasses together and start cooking. When the 4 minute hourglass runs out, turn it over. When the 7 minute hourglass runs out, turn the 4 minute hourglass over. It should have 3 minutes of sand in it, so let it all run out to make up the total of 10 (7 + 3) minutes of cooking time.

Ratio calculations

PAGE 160

1 a 2 : 15 **b** 6 : 11
c 1 : 8 **d** 3 : 5
e 3 : 2 **f** 7 : 1
g 9 : 2 **h** 5 : 7
i 4 : 1 **j** 10 : 1
k 1 : 3 **l** 7 : 11
m 5 : 6 : 10 **n** 5 : 2 : 1
o 61 : 9 : 14
2 27 **3** B
4 a 1 : 4 **b** 3 : 8
c 2 : 5 **d** 1 : 50
e 12 : 7 **f** 2 : 5
g 6 : 25 **h** 3 : 4
i 1 : 5 **j** 18 : 1
k 200 : 39 **l** 5 : 1
5 A **6** 75 min
7 4 : 5 **8** B
9 $105 **10** 2695 mL
11 16 **12** B
13 25 kg **14** 525 mL
15 20 cm **16** $18 288 000

Rate problems

PAGE 162

1 a 88 **b** 51
c 85 **d** 54
e 9.1 **f** 144
g 58 **h** 23
2 a persons/km^2 **b** $/kg
c L/100 km **d** cents/word
e Mb/s **f** $/min
3 $4.20/kg **4** 17/h
5 a 35 700 mL **b** $71.97
6 12.5 min
7 a 364.5 km **b** 47.25 km
8 13.5 sheep/hour **9** Plumber
10 1924 people/year **11** 3.75 runs/over

12 5.2 L/min
13 a 44c/30 s b 1.5c/s
14 $23.84/h
15 a 140 L b 6 weeks
16 25.5 min
17 a 534 m b 225 seconds (3 min 45 s)
18 28.69 cents/kWh
19 8100 m/s
20 $64.93
21 54 min
22 9.58 s

Time calculations

PAGE 164

1 a 0630 b 1620
c 2305 d 1445
e 1918 f 1056
2 a 7:05 a.m. b 6:55 p.m.
c 12:30 p.m. d 8:30 p.m.
e 3:44 a.m. f 3:17 p.m.
3 a 9:00 p.m. b 1:00 p.m.
c 0700 d 3:00 p.m.
e 0900 f 8:00 a.m.
g 10:30 p.m. h 0540
i 2:30 a.m.
4 a 9 h 30 min b 4 h 18 min
c 12 h 36 min d 7 h 24 min
e 15 h 42 min f 10 h 48 min
g 5 h 15 min h 8 h 18 min
5 a 6 h 18 min b 3 h 56 min
c 2 h 53 min d 11 h 25 min
e 7 h 7 min f 6 h 44 min
6 a 4:35 a.m. b 1606
c 10:14 p.m. d 8:25 a.m.
7 a 8h b 8 h 40 min
c 5 h 20 min d 7 h 40 min
e 5 h 13 min f 10 h 43 min
8 a 3:40 a.m. b 12:41 p.m.
c 9:19 a.m. d 10:06 a.m.
e 0106 f 1121
9 a 3:30 p.m. b 3:00 p.m.
c 1:30 p.m. d 3:30 p.m.
e 3:30 p.m. f 3:00 p.m.
10 a 4:00 a.m. b 8:30 p.m.
c 1:45 p.m. d 10:47 a.m.
e 1535 f 0720

9780170454452

Chapter 1

Integers 1 PAGE 08

Part A

1 19 **2** 168
3 An angle that is less than 90° **4** 36 cm^2
5 32 m **6** quadrilateral
7 21, 28, 35 **8** 180

Part B

1 a 3 **b** 115
c 4 **d** 6
2 a 23 **b** 5
3 a $<$ **b** $>$

Part C

1
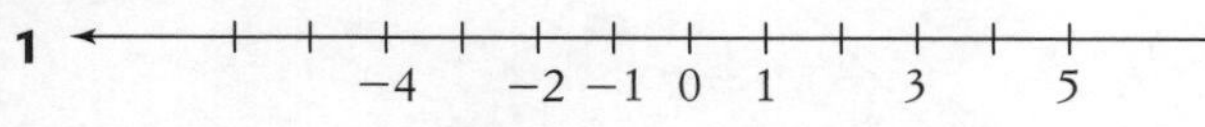

2 a $<$ **b** $>$ **c** $<$
3 a 1 **b** −3 **c** −3

Part D

1 a 8, 4, −2, −3, −6 **b** 1
2 larger (or greater or bigger)
3 a Losing $50 **b** 8 degrees below zero
4 12 steps forward
5 0
6 Because it is not a whole number

Integers 2 PAGE 10

Part A

1 20
2 3.2
3 An angle more than 90° and less than 180°
4 10:05 p.m.
5 25 cm
6 39
7 a
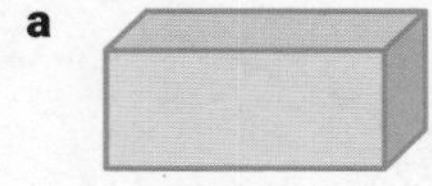
b 6

Part B

1 −7 −6 −5 −4 −3 −2 −1 0 1 2 3 4
2 a $>$ **b** $<$
3 17, 2, 0, −2, −5, −8
4 a −7 **b** −13
5 −12

Part C

1 a −2 **b** −7 **c** 12
2 a −3, −1, 0, 2, 4 **b** 2
3 a −8 **b** 4
4 30 metres below sea level

Part D

1 4°C
2 a 5 **b** 2
3 Teacher to check
4 a Paris **b** 30°C
c 6°C **d** 10°C

Chapter 2

Angles PAGE 20

Part A

1 28.81 **2** 18.4 cm
3 2 **4** 2
5 17 **6** 11:35 a.m.
7 3 **8** 7

Part B

1
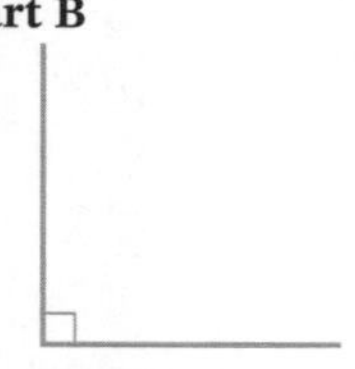

2 right angle
3 a 180 **b** 360
4 an angle less than 90°
5 protractor
6 122°
7 teacher to check

Part C

1
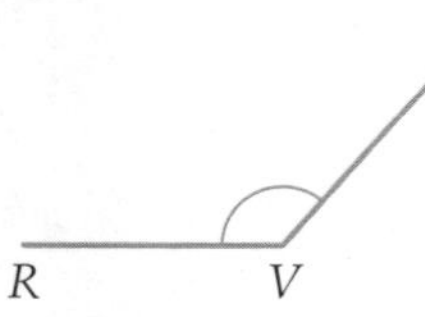

2 a ∠*DBC* or ∠*CBD* **b** ∠*DBA* or ∠*ABD*
c 75° **d** *B*
3 Teacher to check
4 An angle between 180° and 360°

Part D

1 a obtuse angle **b** arms, vertex **c** 154°
d 64° **e** 26°
2 a Teacher to check, e.g. 50° and 60°
b Teacher to check, e.g. 50° and 25°

Angle geometry PAGE 22

Part A

1 a
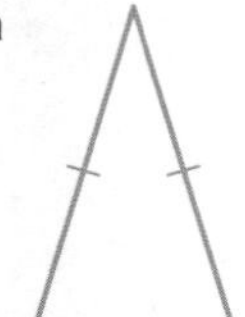
b 1

2 1450
3 192
4 8
5 square pyramid
6 $9
7
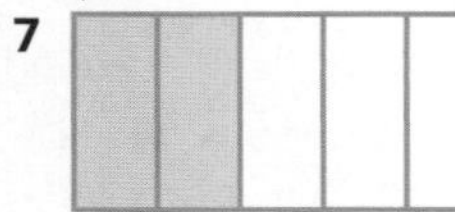

Part B

1 **a** $\angle ADC$ **b** reflex **c** 200° to 250°

2

3 **a** obtuse angle **b** straight angle

4

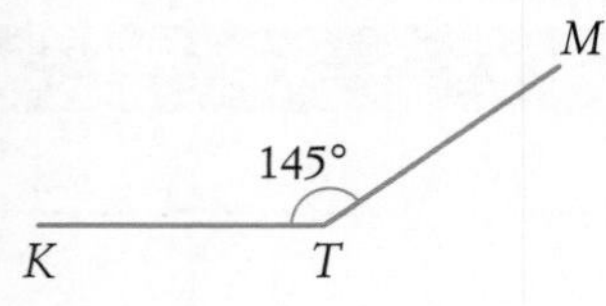

Part C

1 **a** *PQ*, *SR* (or *PS*, *QR*): other answers possible

b *PR*, *SQ*: other answers possible

2 **a** $r = 26$ **b** $y = 65$

c $w = 85$ **d** $g = 89$

3 Teacher to check

4 ||

Part D

1

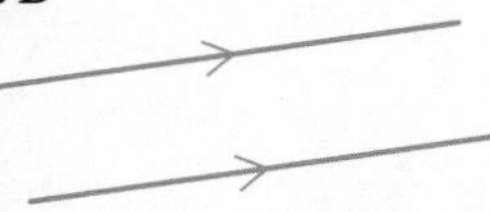

2 e.g. road lanes, railway tracks, sides of a doorway

3 $x = 180 - 108 = 72$ (angles in a straight angle);
$y = 108$ (vertically opposite angles)

4 they add up to 90°

5 right angle

6 Lines that cross at right angles (90°)

7 $a + b + c = 360°$

Angles on parallel lines

PAGE 25

Part A

1 $\frac{1}{2}$

2 108

3 a shape with 8 straight sides

4 10:20 p.m.

5 6.85

6 −21

7 32 m^2

8 $\frac{1}{2}$

Part B

1 15°

2 **a** $x = 27$ **b** $r = 54$ **c** $f = 30$

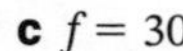

3 **a**

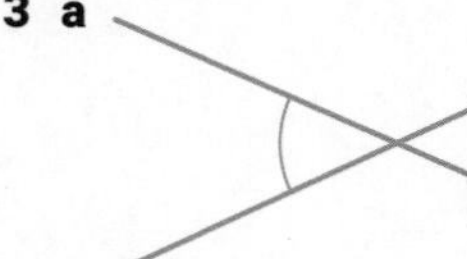

b they are equal

4 **a**

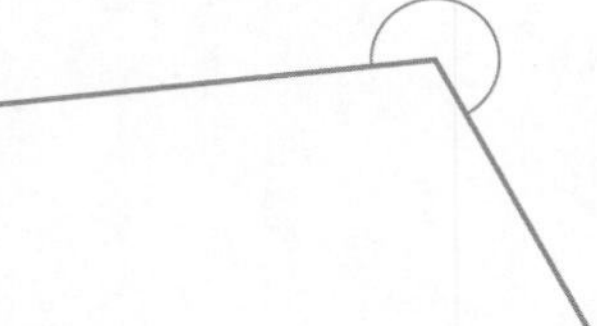

b 149°

Part C

1 **a** transversal

b

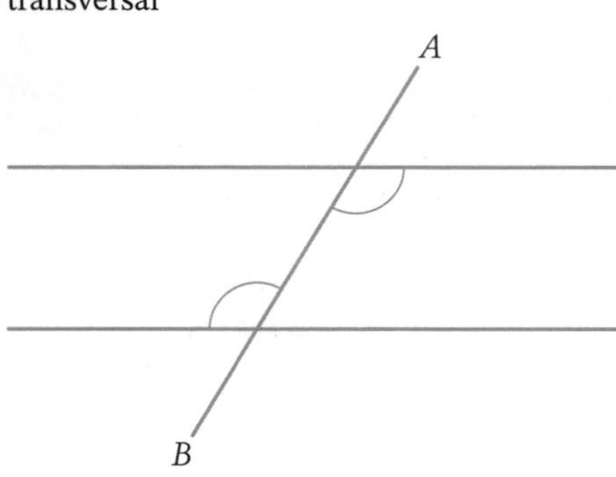

2 $a = 115$, $b = 65$, $c = 115$

3

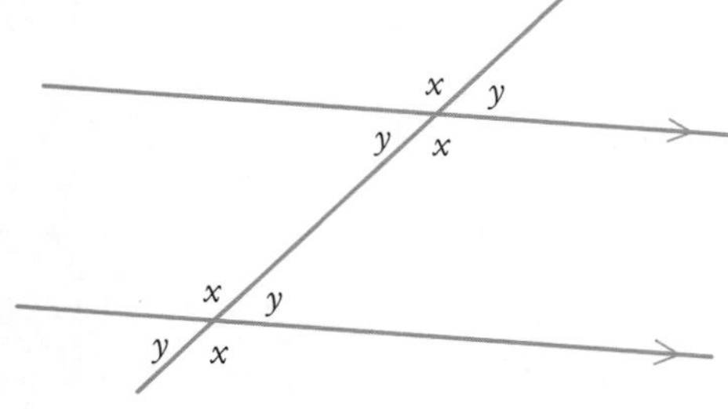

Part D

1 **a**

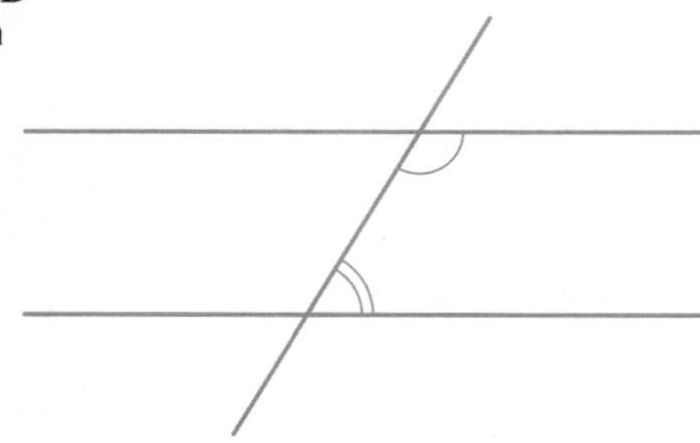

b 2

c Angles between parallel lines on the same side of the transversal

d on (or between), lines, supplementary

2 **a**

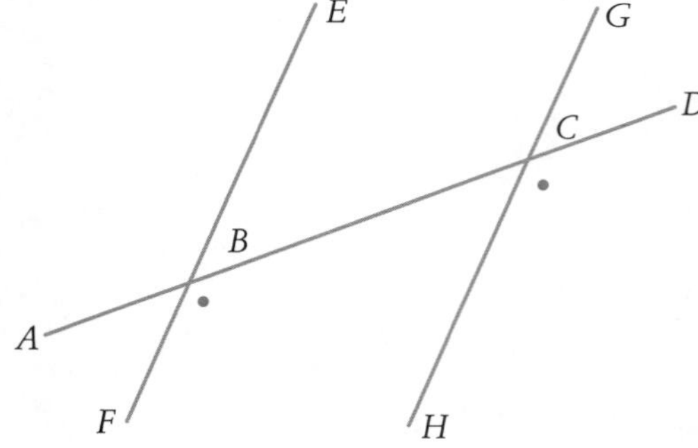

b $\angle DCH$

3 $m = 56$ (alternate angles on parallel lines);
$p = 180 - 56 = 124$ (angles in a straight angle)

Angles revision

PAGE 27

Part A

1 12.4

2 3.42

3 A quadrilateral with all angles 90°

4 11

5 1345

9780170454452

6 $10\ m^2$
7 7
8 14

Part B

1 a $a = 74$ b $m = 68$
2 Teacher to check
3 a $n = 132$ (corresponding angles on parallel lines)
b $d = 90 - 26 = 64$ (angles in a right angle)
4 Teacher to check

Part C

1 supplementary
2 a trapezium b cointerior c 36°
3 $a = 134$, $b = 46$
4 no, alternate angles (105° and 115°) are not equal

Part D

1 a two angles that add up to 90°
b Teacher to check
2 a

b 4
c they are equal
3 $d = 360 - 106 - 84 - 90 = 80$ (angles at a point)
4 reflex angle

Chapter 3

Whole numbers PAGE 36

Part A

1 $\frac{3}{5}$ 2 $12.35
3 1, 5, 25 4 5:00 p.m.
5 81 6 9350
7 26.5 m 8 80

Part B

1 a 48 b 980 c 740
2 Yes
3 a 592 b 25 200 c 13 d 27

Part C

1 a 209 420 b 209 000
2 Teacher to check, exact answer = 454
3 a Yes b No c No
4 22
5 15

Part D

1 14
2 a $16 \times 5 = 8 \times 10 = 80$; halve the first number and multiply by 10 (add a 0)
b 2800 (add two 0s)
3 a 3
b Next digit is 7, which is more than 5 c 16 400
4 a No, last digit is not 0 or 5
b Yes, sum of digits is 18, which is divisible by 9

Powers and square roots PAGE 38

Part A

1 8, 16, 24, 32, 40 2 acute
3 32 4 $15.6\ m^2$
5 4:15 a.m. 6 109
7 $\frac{7}{10}$ 8 412

Part B

1 3 750 000
2 Yes, divisible by 2 (last digit, 4, even) and 3 (sum of digits = 15 divisible by 3)
3 a 121 b 1800 c 62 d 68
4 Teacher to check, exact answer = 864

Part C

1 a 3^5 b $5^2 \times 7^4$
2 a 36 b 512
c 9 d 10
3 a $5 \times 5 \times 5 \times 5$ b 625

Part D

1 base, power (or index)
2 $12^5 = 12 \times 12 \times 12 \times 12 \times 12$
3 a 10 000 000 b 7
4 a $\sqrt{441}$ is the (positive) number which if squared gives the answer 441
b $\sqrt{441} = 21$
5 64

Prime and composite numbers PAGE 40

Part A

1 hexagon 2 90, 180
3 1734 4 3.14
5 31 6 98
7 6800 8 14 m

Part B

1 1100
2 1, 2, 3, 5, 6, 10, 15, 30
3 Teacher to check, e.g. 6 and 8
4 1 5 No
6 1, 11 7 Teacher to check, e.g. 3 and 7

Part C

1 41, 43, 47
2 Teacher to check, $84 = 2^2 \times 3 \times 7$
3 Teacher to check, $60 = 2^2 \times 3 \times 5$
4 a 12 b 420

Part D

1 Multiplied together the common prime factors of 84 and 60: $2 \times 2 \times 3$.
2 A number with more than 2 distinct factors
3 1 4 1, itself
5 Greatest common divisor
6 24 7 72

Whole numbers review PAGE 42

Part A

1 48 **2** 3600
3 13 **4** 7 cm
5 Six hundred and forty-three thousand, five hundred and eighty-nine
6 scalene **7** 45
8 9

Part B

1 $3^5 \times 4^3$ **2** 2401
3 80 **4** 23, 29, 31, 37
5 13
6 Divisible by 6 because divisible by 2 (last digit, 4, even) and 3 (sum of digits = 15, divisible by 3)
7 4

Part C

1 5832
2 5632
3 Last digit, 4, is not 0 or 5
4 Teacher to check, $72 = 2^3 \times 3^2$
5 24
6 Any number from 15 500 to 16 499

Part D

1 composite
2 a 'eight cubed' **b** $8 \times 8 \times 8$ **c** 8
3 a Teacher to check; $28 \times 9 \approx 30 \times 10 = 300$
b $28 \times 9 = 28 \times 10 - 28 = 280 - 28 = 252$
4 square root, 7, 49

Chapter 4

Fractions 1 PAGE 50

Part A

1 obtuse **2** 28
3 −5, −2, 0, 1, 7 **4** 9410
5 212 **6**

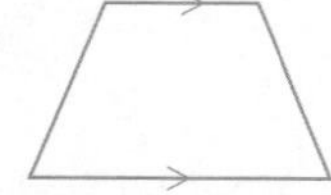

7 1, 4, 9, 16 **8** 22

Part B

1 a 2 squares shaded **b** 3 squares shaded
2 a 6 **b** 40
3 a 3 squares shaded **b** 6 squares shaded
4 a 6 **b** 4

Part C

1 the denominator **2** $\frac{3}{5}$
3 $\frac{14}{3}$
4 a 9 **b** 42
5 a $\frac{7}{12}$ **b** $2\frac{2}{5}$
6 $\frac{2}{5}, \frac{6}{10}, \frac{3}{4}$

Part D

1 a A fraction whose numerator is less than its denominator
b For example, $\frac{2}{5}$
2 is less than
3 a highest/largest, lowest/smallest
b dividing, number/factor
4 mixed numeral
5 Divide 23 by 9 and write the remainder as a fraction with denominator 9: so $\frac{23}{9} = 2\frac{5}{9}$.

Fractions 2 PAGE 52

Part A

1 36 **2** 5
3 9:24 p.m. **4** 20 m
5 9 **6** 9600
7 **8** 54

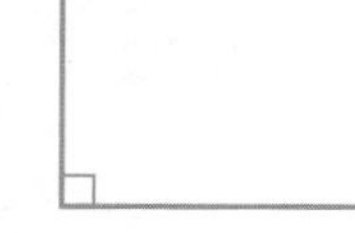

Part B

1 $\frac{5}{7}$ **2** 8 rectangles shaded
3 $\frac{9}{20}$ **4** $2\frac{3}{7}$
5 $\frac{1}{3}$ **6** $\frac{3}{4}$
7 Teacher to check, e.g. $\frac{7}{28}$ **8** 6, 12, 18, 24

Part C

1 a $\frac{1}{3}$ **b** $1\frac{1}{3}$
c $\frac{9}{20}$
2 \$48
3 a $\frac{1}{3}$ **b** $\frac{8}{15}$
c $5\frac{7}{8}$
4 8 months

Part D

1 a \$15 **b** \$5
c $\frac{1}{3}$
2 a $2\frac{7}{12}$ **b** 2
3 A fraction whose numerator is greater than or equal to its denominator
4 numerator **5** denominator

Fractions 3 PAGE 54

Part A

1 6 **2** parallelogram
3 1, 2, 4, 5, 10, 20 **4** \$10
5 48 m **6** 120
7 0325 **8** 273

9780170454452

Part B

1 $\frac{22}{25}$

2 Yes, $\frac{7}{8}=\frac{21}{24}$ and $\frac{5}{6}=\frac{20}{24}$ so $\frac{21}{24}>\frac{20}{24}$

3 a $7\frac{3}{20}$ b $4\frac{5}{8}$

c $13\frac{11}{18}$

4 $\frac{7}{8}$

5 2 h 40 min

6 7.5

Part C

1 $\frac{35}{4}$

2 Answer is always 1, e.g. $\frac{2}{3}\times\frac{3}{2}=\frac{6}{6}=1$

3 a $\frac{1}{8}$ b $3\frac{3}{4}$

c $\frac{5}{6}$ d $1\frac{1}{2}$

e $\frac{1}{81}$

4 $\frac{7}{10}$

Part D

1 $4\frac{9}{10}$

2 reciprocal

3 $\frac{4}{15}$

4 Teacher to check; $\frac{9\times 2}{10\times 3}=\frac{18}{30}=\frac{3}{5}$ or $\frac{3}{5}\times\frac{1}{1}=\frac{3}{5}$ after reducing

5 a 5 b 52

c $\frac{13}{20}$

Percentages

PAGE 56

Part A

1 8.109

2 −1

3 4

4 90°

5 45 600

6 7

7 9, 18, 27, 36, 45

8 7

Part B

1 $\frac{1}{15}$

2 $0.80

3 240

4 a $\frac{1}{4}$ b $\frac{1}{10}$

5 $45

6 a 75% b 20%

Part C

1 a $\frac{33}{50}$ b 0.66

2 $180

3 15%

4 $\frac{1}{8}$

5 a 40% b 1.5%

6 84

Part D

1 $\frac{1}{3}$, because $\frac{1}{3}=\frac{10}{30}$ while 30% $=\frac{3}{10}=\frac{9}{30}$

2 85%

3 denominator, 100

4 a $18 b $102

5 Multiply it by 100%

6 a 15% b $\frac{17}{20}$

Chapter 5

Mental calculation

PAGE 64

Part A

1 600

2 $\frac{7}{10}$

3 12 cm

4 −7, −3, 0, 5, 6

5 4650

6 Teacher to check

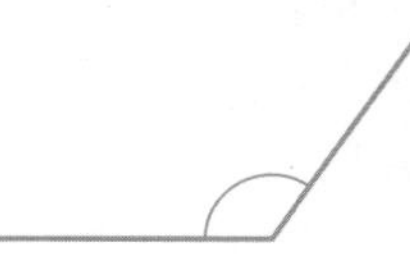

7 1, 2, 3, 4, 6, 12

8 4

Part B

1 15

2 a 66 b 66

3 a true b false

c true d false

4 11

Part C

1 a 3600 b 11

c 92 d 90

2 a $25\times 10-25\times 2$ b $17\times 10+17\times 1$

3 a true b false

Part D

1 a $3\times 8\times 10=24\times 10=240$

b $7\times(5\times 4)=7\times 20=140$

c $(63+7)+(18+12)=70+30=100$

2 a commutative, multiplication

b $ab+ac$

3 $37\times(10-1)=37\times 10-37\times 1=370-37=333$

Algebra

PAGE 66

Part A

1 20

2 360

3 80

4 13

5 300

6

7 $2+3+5=10$

8 58

Part B

1 a 1600 b 1800
c 385

2 a 5 b 9
c 50 d 6

3 $ab - ac$

Part C

1 a $3ab$ b $3r$
c $9 - 3d$

2 a $n + 4$ or $4 + n$ b $3n - 6$

3 a 13 b 3
c 5

Part D

1 a $\$20 - \$p$ b $\frac{18}{n}$

2 $-x \times (-x) = x^2$

3 a 4 times w times w
b y plus 5, then divided by 2

4 1, any number, except zero, divided by itself equals 1

5 substitute 6 77°F

Equations 1 PAGE 68

Part A

1 4 2 1:50 pm

3 4, 8, 12, 16, 20 4 180

5 \$23 6 0.674

7 144 8 4 cm

Part B

1 a b b $2w^2$
c $\frac{10}{de}$

2 a $x - 3y$ b $\sqrt{2t}$
c $\frac{a+b}{2}$ or $\frac{1}{2}(a + b)$

3 a 3 b -13

Part C

1 $q = 7\frac{1}{2}$ 2 $d = -40$

3 $z = -1$ 4 $r = 0$

5 $n = 8$ 6 $e = 3$

7 $n = 44$ 8 $y = 15$

Part D

1 $n = 7$

2 variable or pronumeral

3 a 5 times a number, c, less 4, equals 11
b Opposite operations, for example, addition and subtraction.
c $c = 3$ d $5 \times 3 - 4 = 11$

Equations 2 PAGE 70

Part A

1 18 2 A triangle with 2 equal sides

3 4 6

A, B, C, D, E

5 3 6 8, 16, 24, 32, 40

7 10 8 2, 3

Part B

1 a $a = 4$ b $y = 32$

2 a $2l + 2w$ b lw

3 a $d + \sqrt{e}$ b $\frac{\$p}{4}$

Part C

1 a $n - 14 = 8, n = 22$ b $2a - 6 = 18, a = 12$

2 a 8 b 0

3 a 72 cm^2 b 7 cm

Part D

1 $x + (-x) = 0$ 2 $k = 9$

3 a solution b formula

4 Any number multiplied by 1 is itself

5 $ab = ba$

6 $\$n - \$25y$

Chapter 6

Transformations and symmetry PAGE 80

Part A

1 25% 2 0 3 $a + b = 90$ 4 $\frac{4}{5}$

5 34 6 head, tail 7 $m = 7$ 8 2×3^3

Part B

1 translation, reflection, rotation

2

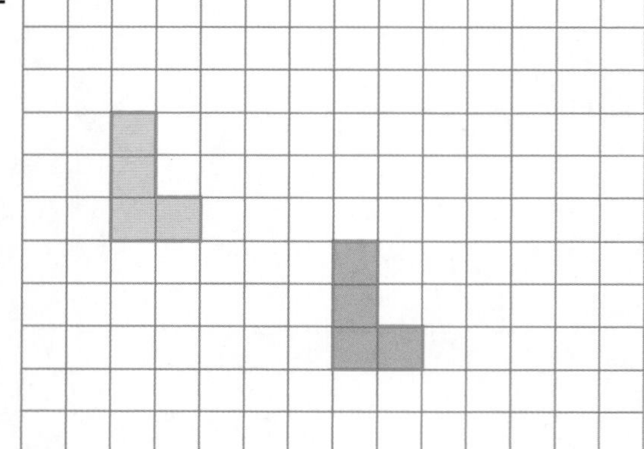

3 a 4 b 5 c 2

4

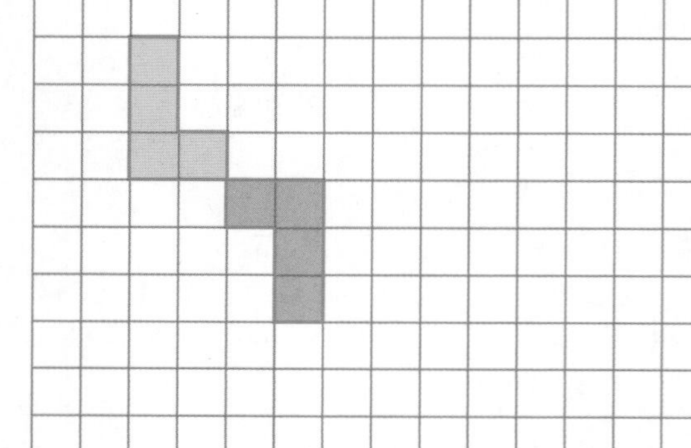

5 reflection

6

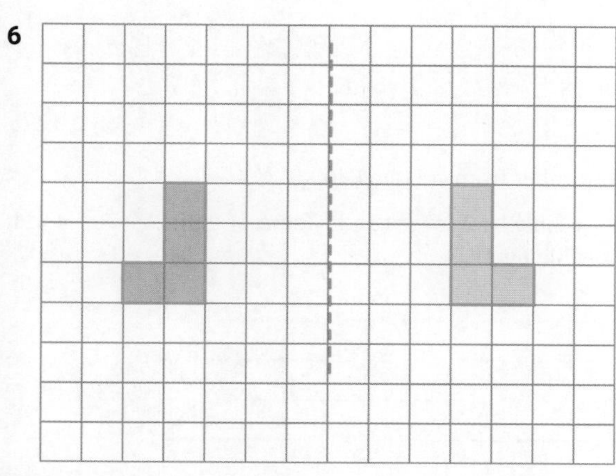

Part C

1

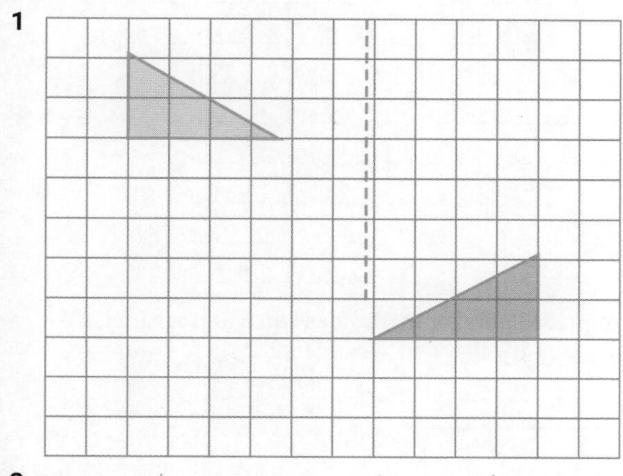

2 **a**

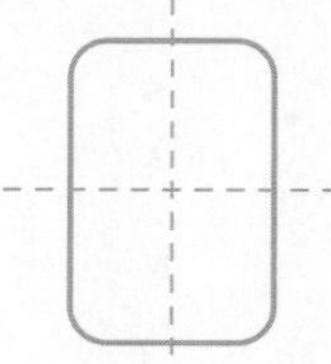

b

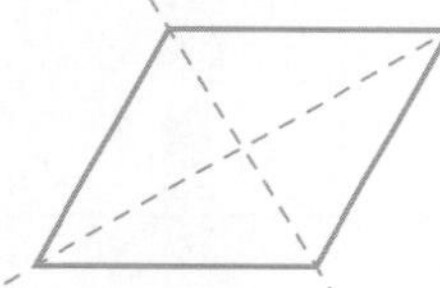

3 Teacher to check, e.g.

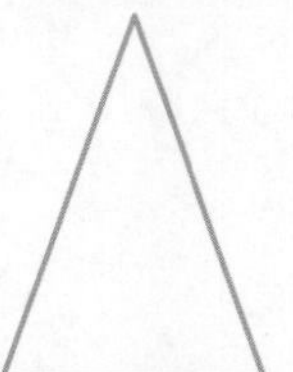

4 You can fold it in half along any line that passes through its centre (any diameter)

5

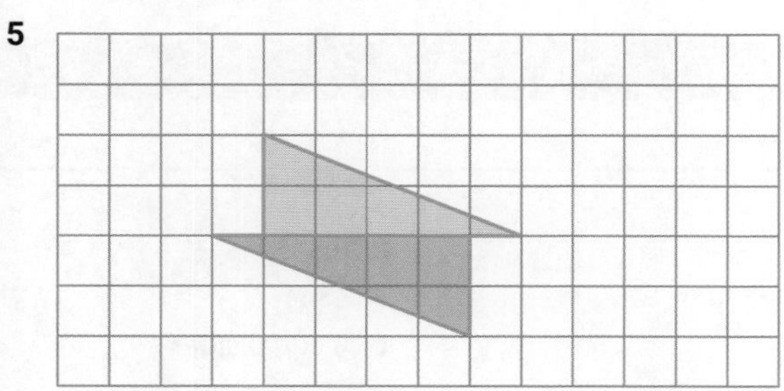

6 **a** 3 **b** 6

Part D

1 image

2 reflection

3 A spin or revolution

4 **a** rotation **b** a translation

5 It will not fit/map onto itself when spun

6 **a** 270° **b** 180°

Triangles

PAGE 83

Part A

1 −14 2 125° 3 20% 4 $\frac{2}{21}$

5 $12ay$ 6 9 7 29.1 8 11:15 a.m.

Part B

1, 2

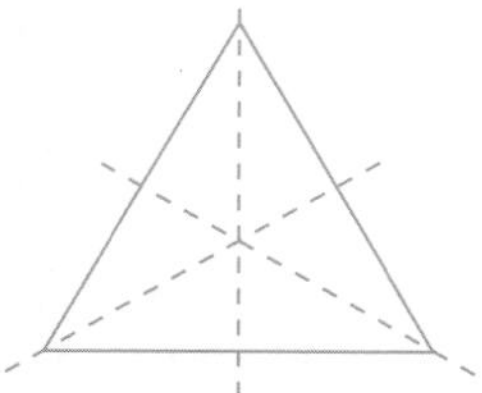

3 3

4 **a** isosceles **b** right-angled

5 **a** scalene **b** isosceles

6

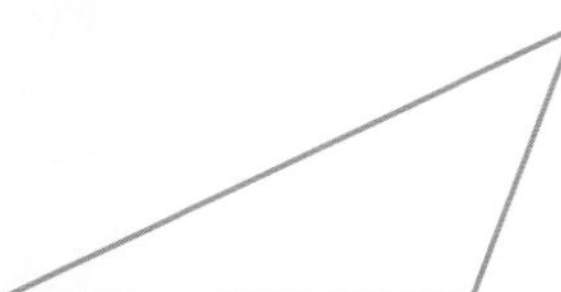

Part C

1 **a** equilateral, acute-angled **b** scalene, right-angled

2

3 **a** $a = 116$ **b** $b = 85$ **c** $c = 102$
d $d = 42$ **e** $e = 48$

Part D

1 acute-angled, obtuse-angled, right-angled

2 180°

3 **a** 60° **b** 120°

4 triangle, sum, opposite

5 90°

6 A triangle with all angles acute (under 90°)

7 exterior

Quadrilaterals

PAGE 86

Part A

1 120

2

3 no

4 62.5%

5 $n = -5$

6 $\frac{3}{14}$

7 8400

8 0

Part B

1

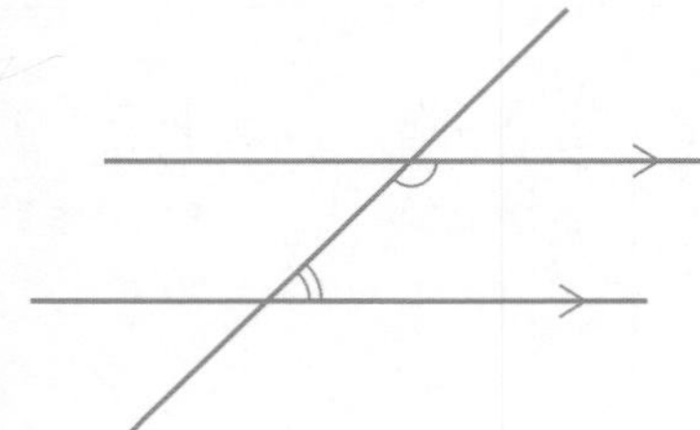

2 $y = 25$

3 isosceles, obtuse-angled

4 **a** square **b** parallelogram **c** scalene triangle **d** pentagon

Part C

1

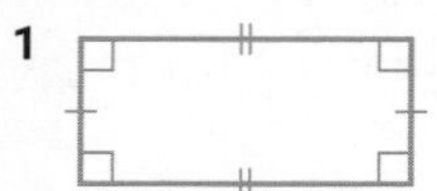

2 $x = 158, y = 22$

3 **a** 0 **b** 1

4

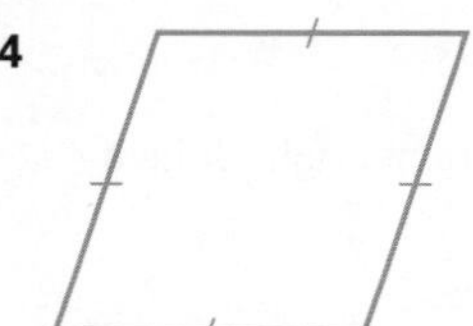

5 2

6 $h = 51$

Part D

1 A shape with 4 straight sides

2 360°

3 A square has all sides equal while a rectangle has opposite sides equal

4 square, rectangle, rhombus, parallelogram

5 **a** rectangle **b** rhombus **c** kite

Geometry revision PAGE 89

Part A

1 10 000

2

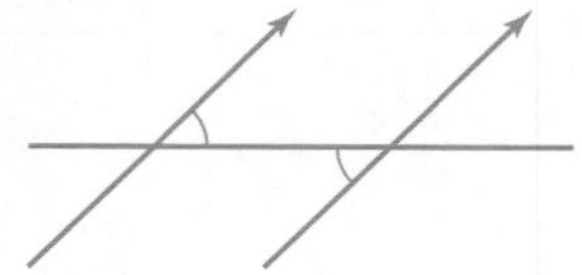

3 \$45

4 **a** $2n - 8$ **b** 12

5 2

6 Any number from 7000 to 8400

7 It is divisible by 2 (even) and 3 (sum of digits $6 + 8 + 4 = 18$ is divisible by 3)

Part B

1

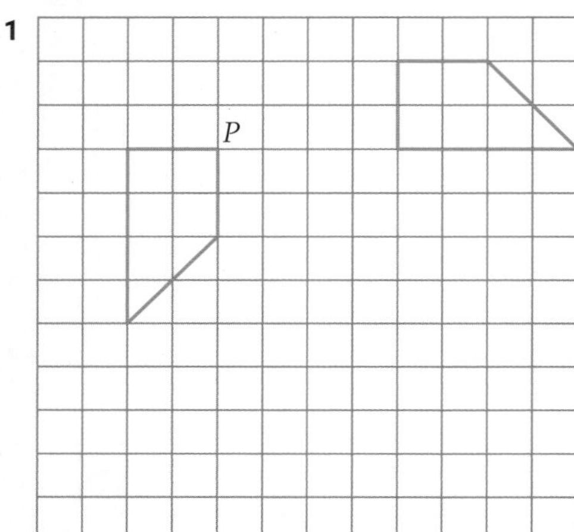

2 **a** A quadrilateral with all sides equal and all angles equal (90°)

b

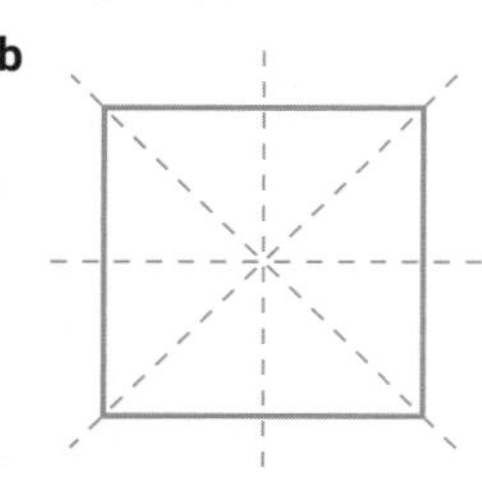

c 4

3 **a** reflection, rotation **b** reflection

Part C

1 **a** isoceles, right-angled **b** 45 **c** 1 **d** 0

2 **a** $x = 62°$ **b** $p = 76, q = 128$ **c** $y = 87$

Part D

1 A quadrilateral with all sides equal

2 All sides equal, all angles equal (60°)

3 **a, b** teacher to check, e.g. angles 40°, 35°

c obtuse-angled

4 **a** kite **b** rectangle

Chapter 7

Decimals 1 PAGE 96

Part A

1 125

2 0.845 L

3 $t = 65, u = 115$

4 6864

5 15

6 3 equal sides

7 32 cm

Part B

1 72

2 640, 66, 63, 60, 6

3 3100

4 180

5 $\frac{3}{25}$

9780170454452

6 a 3980 **b** 4000

7 $\frac{49}{200}$

Part C

1 0.007

2 a 13.12 **b** 22.27

c 127.58 **d** 6.75

3 Teacher to check, e.g. 4.17

4 a 8.16 **b** 3.21

Part D

1 a 3 **b** 0

2 a Jay **b** 0.58 m

c 8.22 m

3 $\frac{36}{100} = \frac{9}{25}$

3 *Chess Made Easy, Ten-pin Bowling for Winners, The Snooker Book, Mah Jong Explained*

Decimals 2

PAGE 98

Part A

1 1800 **2** 1835

3 scalene, acute-angled **4** 64

5 $4r$ **6** 6

7 300

Part B

1 320 **2** 0.32

3 21.76 **4** 7.04, 7.3, 7.35, 7.47, 7.8

5 $\frac{13}{20}$

6 a 84 000 **b** 84

7 32.15

Part C

1 a 456 200 **b** 0.864

c 1629.6 **d** 0.0028

e 0.174 **f** 13.69

g 10.8025 **h** 4740

Part D

1 one place to the left

2 a 4.7 s **b** 190.5 s

c 38.1 s

3 a 3.341 **b** 33.41

4 whole, right, 416.8 ÷ 8, 52.1

Decimals 3

PAGE 100

Part A

1 4 **2** 4

3 13, 17 **4** triangular pyramid

5 41 **6** 62.8

7 21 600 **8** 3 m

Part B

1 10.061, 10.07, 10.3, 10.6, 10.63

2 $\frac{19}{50}$ **3** 300.6

4 409.5, 22.75 **5** 15

6 3.089

7 a 4.84 **b** 0.0484

Part C

1 a 0.375 **b** 0.83

2 36.8

3 a 41.6 **b** 41.638

4 $11.85 **5** 5.95 m

6 Any decimal from 4.085 to below 4.09

Part D

1 a recurring **b** 0.272 727 ...

2 7, 4, down, less, 195.7

3 a $0.35 **b** $555

c $903.80

Decimals revision

PAGE 102

Part A

1 18 **2** 30

3 4, 6, 8, 9 **4** $y = 130$

5 $\frac{2}{7}$ **6** 58 200

7 $n = 14$ **8** 40 m

Part B

1 3.46, 3.6, 3.645, 3.69, 3.7 **2** 14.03

3 a 34 610 **b** 0.062 58

c 330.38 **d** 532.48

e 0.708 **f** 566.05

Part C

1 a 0.3125 **b** $0.\dot{5}7142\dot{8}$

2 7.7 **3** 0.512

4 Any decimal from 25.05 to below 25.15

5 a $9.25 **b** $24.10

c $25.90

Part D

1 A terminating decimal has digits that end while a recurring decimal has digits that repeat endlessly.

2 $0.3\dot{1}\dot{6}$ **3** 3.22

4 3 **5** $542.30

6 48.08 **7** 39.7 L

Chapter 8

Measurement

PAGE 113

Part A

1 −12 **2** reflection

3 $\frac{4}{5}$ **4** $w = 72$

5 600 **6** $27a^2b$

7 $1 + 4 + 9 + 16 = 30$ **8** $\frac{1}{20}$

Part B

1 a 5100 **b** 780

c 4 **d** 0.547

2 12 cm **3** $x = 4, y = 3$

4 square centimetre (cm^2)

Part C

1 a 22.6 cm **b** 26.86 cm^2

2 a 1 000 000 **b** 100

3 30.6 m

4 **a** 14 m^2 **b** 0.25 m^2
c 56

Part D

1 A unit of area equal to the area of a square of length 1 m

2 Perimeter of a rectangle

3 one-thousandth $\left(\frac{1}{1000}\right)$

4 Mega

5 cubic metres

6 perimeter

7 $A = lw$, where A is area, l = length, w = width.

Area PAGE 115

Part A

1 3500 **2** 600

3 $a = 135, b = 135$ **4** -1

5 $c = 14$ **6** 5600

7 

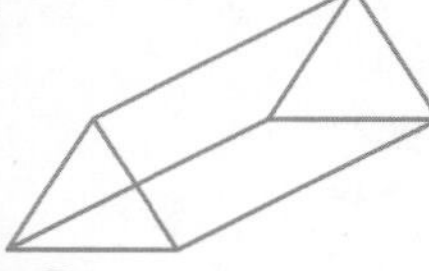

Part B

1 **a** 48 m **b** 12 cm

2 0.64

3 **a** 4.5 m **b** 20.25 m^2

4 **a** 59.4 m^2 **b** 3 m^2 or 30 000 cm^2

5 10 000

Part C

1 **a** 16 m^2 **b** 13.5 cm^2
c 33 m^2 **d** 40.5 cm^2
e 44 cm^2 **f** 30.4 m^2
g 25 cm^2 **h** 28 m^2

Part D

1 hectare

2 **a** area of a triangle
b A = area, b = base length, h = perpendicular height

3 **a** 

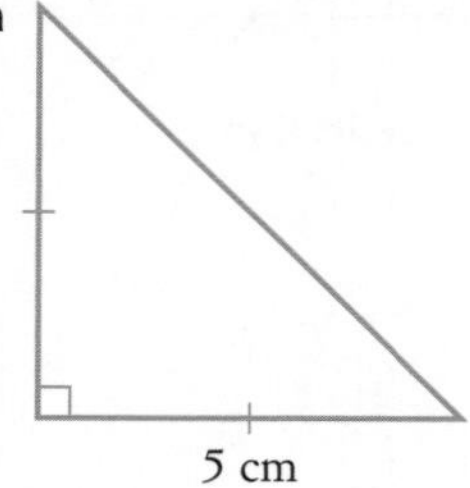

b 12.5 cm^2

4 **a** $A = bh$
b multiply the base length by the perpendicular height

5 any two values that have a product of 40, e.g. 8 and 5

Volume and capacity PAGE 118

Part A

1 81 **2** 39

3 63 **4** $n = 70$

5 $\frac{7}{8}$ **6** 3600

7 3 **8** 4

Part B

1 **a** 32 m **b** 54 m^2

2 7.5 cm **3** 10 000

4 10 000 **5** 5 m

6 **a** 6 cm^2 **b** 18 m^2

Part C

1 1 000 000 cm^3

2 0.001 L

3 **a** 108 m^3 **b** 56.448 cm^3

4 4 m

5 504

6 **a** 94 500 cm^3 **b** 0.0945 m^3

Part D

1 A unit of volume equal to the volume of a cube with length 1 cm

2 prism

3 **a**

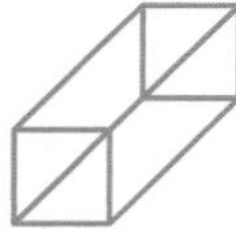

b A solid with identical square cross-sections

4 millilitre

5 3.42 m^3

6 **a** Volume of a rectangular prism
b V = volume, l = length, w = width, h = height

Chapter 9

The number plane 1 PAGE 126

Part A

1 16.8 **2** $m = 35$ **3** 3^5

4 **a**

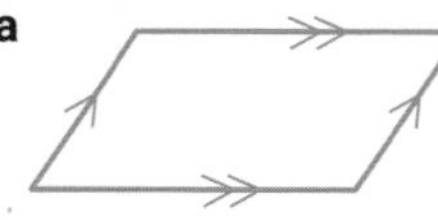

b 2

5 9 **6** $12k$ **7** 22 m^2

Part B

1 **a** sports oval **b** church or railway crossing

2 **a** E2 **b** C1 **3** (3, 2)

4 y, 0, 1, 2, 3, 4, 5, x, A, C

5 (2, 8), (2, 2), (5, 5) or (−1, 5)

9780170454452

Part C

1 a $(-3, 2)$ **b** $(-4, -2)$

2 3rd

3

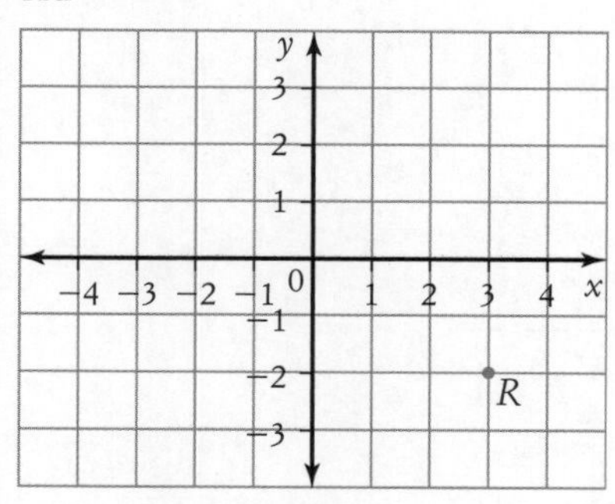

4 a

b trapezium **c** *WZ* and *XY*

Part D

1 map, spreadsheet, theatre seating

2 a vertical axis or reference line going up and down

b the point (0, 0) or where the *x*- and *y*-axes cross

3 0

4 negative, positive

5 From the origin (0, 0), go 2 units to the right, then go 4 units down to mark the point with coordinates $(2, -4)$.

6 *x*-axis.

The number plane 2

PAGE 128

Part A

1 0.032 **2** square **3** 72 **4** -4

5 $12ab$ **6** 33 m^2 **7** 625 **8** $c = 7$

Part B

1 a (0, 0)

b (3, 0) or any point of the form $(a, 0)$

c (4, 5) or any point with two positive coordinates

d $(2, -6)$ or any point with a positive *x*-coordinate and a negative *y*-coordinate

2

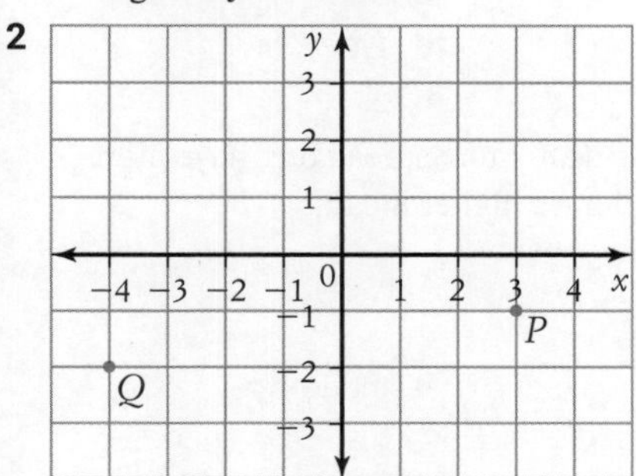

3 a $(-3, 2)$ **b** $(0, -3)$

Part C

1 a

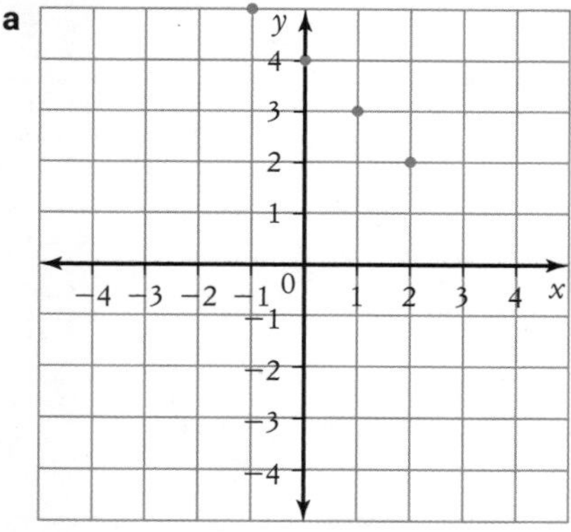

b They lie on a straight line.

c (3, 1) or any point whose coordinates add to 4

2 a right-angled (or scalene)

b

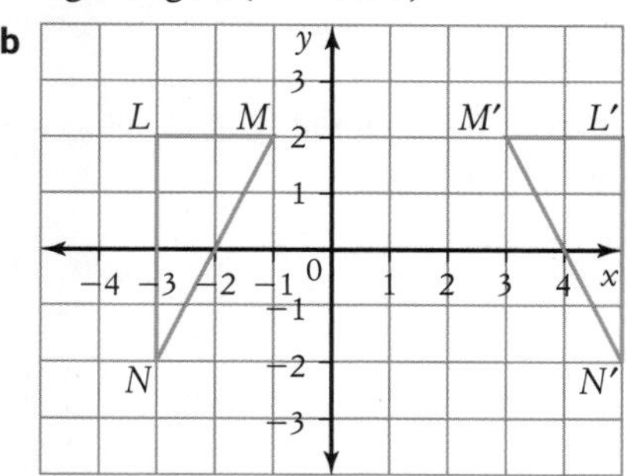

c $N(-3, -2)$, $N'(5, -2)$

Part D

1 Reflection across the *y*-axis

2 a

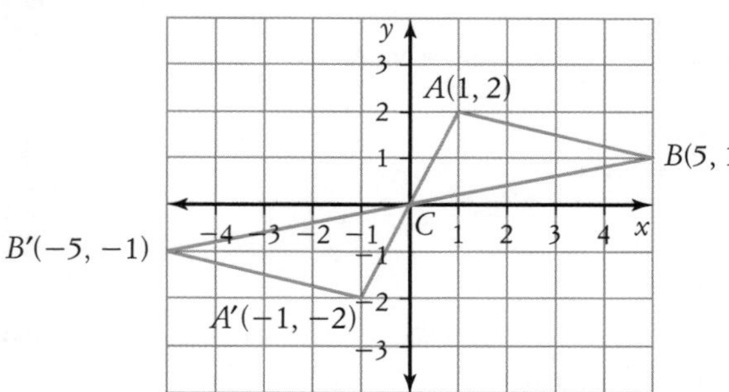

b $A'(-1, -2)$, $B'(-5, -1)$

3 $3y$

4 a

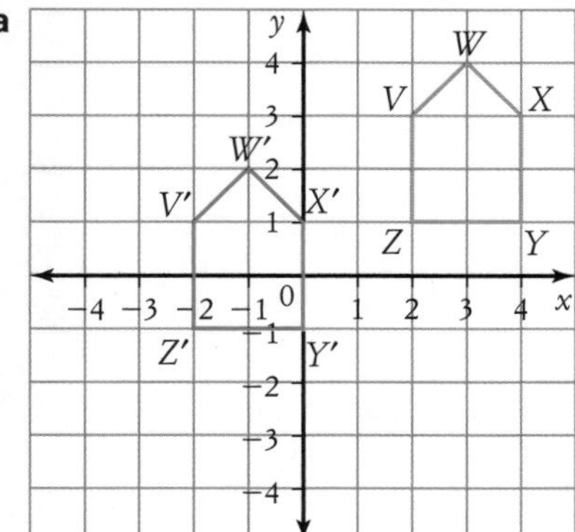

b The coordinates for *Z* are (2, 1) and $(-2, -1)$ for *Z*′, so the coordinates for *Z*′ are the inverse of *Z*.

Chapter 10

Statistics 1 PAGE 136

Part A

1 $6b$ **2** 75
3 Teacher to check **4** 2028
5 No **6** 0.03
7 360° **8** 23 cm

Part B

1 a column graph **b** days of the week
c 2 hours
d Friday; no school/work the next day, so he can stay up late
e 4
f Tuesday
g 4.2 hours

Part C

1 a The temperature scale does not start at zero and is uneven between 26°C and 30°C.
b 30°C **c** 3 p.m.
d 11 a.m.

2 a 2 3 4 5 6 7 8 9 **b** 9

c 8 **d** $\frac{7}{12}$

Part D

1 a sector graph (or pie chart or circle graph)
b sleep
c $\frac{1}{8}$
d meals, homework
e 360
f 105°
g TV + meals (or homework) + exercise (or computer)
h It shows clearly the parts of a whole.

Statistics 2 PAGE 138

Part A

1 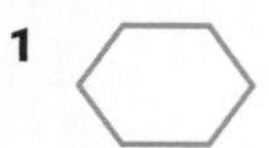

2 $0.1\dot{6}$ **3** 26
4 $y - 10$ **5** rectangle (or square)
6 $2 \times 3^2 \times 5$ **7** $\frac{3}{10}$
8 $p = 69$

Part B

1 a 15, 16, 17, 18, 20, 22, 22; middle temp = 18°C
b 18.6°C
c 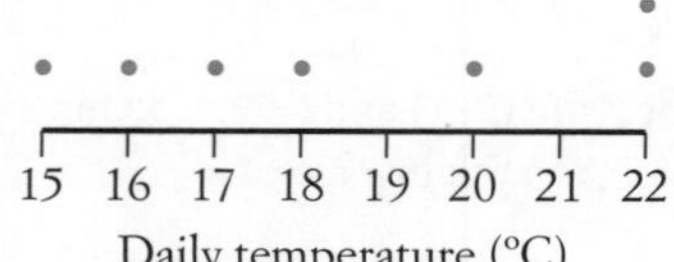

d 22°C **e** Sunday
f 3 **g** 7°C

Part C

1 a 20 **b** 45
c 35 **d** $\frac{1}{2}$

2 a 33 **b** 69
c 35.3
d

Stem	Leaf
2	2 4 6 6 6
3	0 3 3 7
4	0 5 8
5	
6	9

Part D

1 a range **b** mode
c mean
2 outlier
3 a 9 **b** 8 or 12
c 9, 10 or 11 **d** 12

Statistics 3 PAGE 140

Part A

1 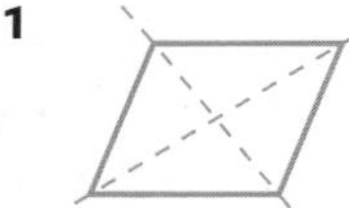 **2** \$19

3 700 **4** False
5 $u = 17$ **6** 0.1932
7 12.3 cm^2

Part B

1 a 186.5 **b** 34
c 170 **d** 188
e 185 **f**

Stem	Leaf
17	0 9
18	3 5 5 8
19	2 5 9
20	4

g $\frac{2}{5}$ **h** decrease

Part C

1 a 2 **b** 2
c 4 **d** 1.7
2 a 7 **b** 2
c The summer scores are more spread out, have more higher values and have higher mean.
d Winter

Part D

1 a range **b** mode
2 mean, range
3 a 23 **b** 20.5
c 23 **d** 20

Statistics revision

PAGE 142

Part A

1 16 **2** $45

3 $x = 48$ **4** 6:05 p.m.

5 1, 2, 4, 5, 10, 20 **6**

7 36 **8** 14

Part B

1 a

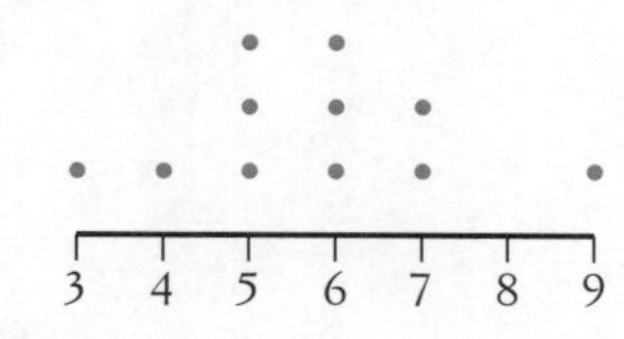

b 9 **c** 5

d 5.5 **e** 5.7

f $\frac{1}{4}$ **g** 6

h mean

Part C

1 a $\frac{2}{3}$ **b** 73

c 79 **d** 72.25

e 42

2 a higher **b** the same

c higher

Part D

1 a back-to-back

b 36 **c** 7S

d 7S's scores are generally higher.

e 69 **f** 84.5

g both classes **h** 70

Chapter 11

Probability 1

PAGE 152

Part A

1 248 **2** 30%

3 0

4

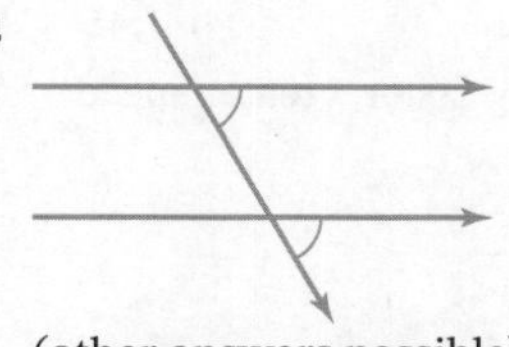

(other answers possible)

5 $3\frac{1}{3}$ **6** 7850

7 a $3x + 6$ **b** $x = 4$

Part B

1 0.85

2 a 3 **b** 8

c 4

3 a impossible **b** even chance

c unlikely

4 spring, summer, autumn, winter

Part C

1 a 2 **b** 12

c 90

2 a {hearts, diamonds, clubs, spades}

b {male, female}

3 $\frac{3}{13}$

4 a $\frac{1}{3}$ **b** $\frac{1}{3}$

Part D

1 Teacher to check, e.g. the sun rose this morning

2 sample space

3 a Each possible outcome (nurse) was equally likely

b 30%

4 $\frac{1}{2}$

5 The probability of an event, E

6 a red, amber, green

b Each outcome/colour is not equally likely; amber appears for a shorter time, red or green may appear for a longer time

Probability 2

PAGE 154

Part A

1 y-axis **2** kite

3 One possible answer is shown below.

4 9.8

5 0.003 49 **6** −2

7 12.5% **8** 15.75 cm^2

Part B

1 $\frac{13}{40}$

2 a {1, 2, 3, 4, 5, 6} **b** $\frac{2}{3}$

3 Improbable means a low/little chance while impossible means no (zero) chance.

4 a 0.25 **b** 0

c 1

Part C

1 Teacher to check, e.g. rolling 7 on a die

2 a 45% **b** 55%

c 100%

3 a 15 **b** 44

4 a $\frac{1}{4}$ **b** 13

Part D

1 a chance of $\frac{1}{2}$; equal chance of happening as not happening

b Teacher to check, e.g. tossing tails on a coin

2 1

3 a low chance, very unlikely **b** 92%

4 a $\frac{3}{4}$ **b** 1

5 0

Chapter 12

Ratios PAGE 166

Part A

1 −3, −1, 0, 4, 5, 8 **2** No
3 Teacher to check **4** 60%
5 $\frac{a+b}{2}$ **6** 0
7 59.62 **8** $d = 90$

Part B

1 $\frac{2}{5}$ **2** 32
3 a $\frac{2}{5}$ **b** 60%
4 180 **5** 2400
6 a $\frac{2}{5}$ **b** $\frac{1}{3}$

Part C

1 a 5 : 3 **b** 3 : 8
2 30 **3** 24
4 a 2 : 5 **b** 5 : 2 : 4
c 9 : 20 **d** 3 : 10

Part D

1 18
2 a One part cordial for every 6 parts of water
b stronger
3 a \$60 **b** 1 : 2
c 2 : 3
4 35 kg
5 1 : 1, equal numbers of boys and girls

Rates PAGE 168

Part A

1 −8, −5, −1, 2, 3, 7
2 a An angle greater than 180° and less than 360°
b 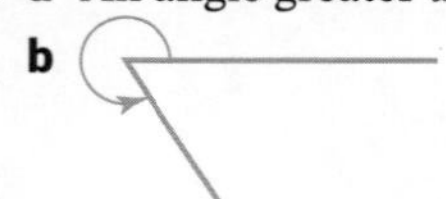

3 Yes **4** $\frac{9}{20}$
5 $2ab - c$
6 $k = 63$ **7** $\frac{5}{26}$

Part B

1 a \$7.10 **b** \$21.30
2 1520c
3 a 7.5 km **b** 37.5 km
4 39
5 a \$15.20 **b** \$121.60

Part C

1 a 8 c/min **b** 22 m/s
c 26 students/teacher
2 B **3** \$26.88
3 a 1440 **b** 104 min (1 h 44 min)
5 C

Part D

1 \$1.85
2 a cents per litre **b** \$77.80
c 40.9 L
3 a km/h (or m/s)
b persons/year
4 8 km/h
5 3 h 7 min

Rates and time PAGE 170

Part A

1 $\frac{1}{6}$ **2** yes
3 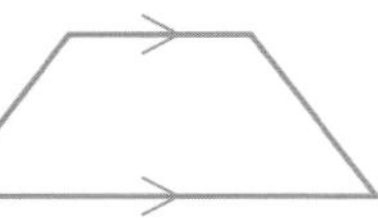

4 −18
5 $n + 2$
6 Any two angles that add to 180°, e.g. 30° and 150°
7 256 **8** $n = 17$

Part B

1 20:35 **2** 2:20 p.m.
3 a 8 **b** 8.5
4 240 **5** 3 h 25 min
6 a 75 L **b** 6 min 40 s (400 s)

Part C

1 5h
2 a 4 h 40 min **b** 6 h 40 min
3 a Sam started returning home
b 12 noon **c** 2 km/h
4 a 10:35 p.m. **b** 4:50 p.m.

Part D

1 No need to write a.m./p.m.; easier for some time calculations
2 a 33 min **b** Stone St. **c** 7:45
3 a distance **b** travelling faster, greater speed
4

Distance from home (km)
12
10
8
6
4
2
9 a.m. 11 a.m. 1 p.m. 3 p.m.
Time

9780170454452